P9-BHX-642

DESTRUCTION OF HAZARDOUS CHEMICALS IN THE LABORATORY

DESTRUCTION OF HAZARDOUS CHEMICALS IN THE LABORATORY

George Lunn
Eric B. Sansone

Program Resources Inc.
Environmental Control and Research Program
NCI-Frederick Cancer Research Facility
P.O. Box B
Frederick, MD 21701

A WILEY-INTERSCIENCE PUBLICATION
JOHN WILEY & SONS, INC.
NEW YORK / CHICHESTER / BRISBANE / TORONTO / SINGAPORE

Extreme care has been taken with the preparation of this work. However, neither the publisher nor the authors warrants the procedures against any safety hazards. Neither the publisher nor the authors shall be held responsible or liable for any damages resulting in connection with or arising from the use of any information in this book.

Copyright © 1990 by John Wiley & Sons, Inc.

All rights reserved. Published simultaneously in Canada.

Reproduction or translation of any part of this work
beyond that permitted by Section 107 or 108 of the
1976 United States Copyright Act without the permission
of the copyright owner is unlawful. Requests for
permission or further information should be addressed to
the Permissions Department, John Wiley & Sons, Inc.

Library of Congress Cataloging in Publication Data:

Lunn, George.
    Destruction of hazardous chemicals in the laboratory / George
Lunn, Eric B. Sansone.
        p.   cm.
    "A Wiley-Interscience publication."
    Includes bibliographical references.
    ISBN 0-471-51063-7
    1. Hazardous wastes—Safety measures. 2. Chemical laboratories—
Safety measures. I. Sansone, E. B. (Eric Brandfon), 1939–
II. Title.
    TD1050.S24L86   1990   66104
    604.7—dc20   89-21533
                                                              CIP

Printed in the United States of America

10 9 8 7 6 5 4 3 2

Preface

This book is a collection of detailed procedures that can be used to degrade and dispose of a wide variety of hazardous chemicals. The procedures are applicable to the amounts of material typically found in the chemical laboratory. Exotic reagents and special apparatus are not required—the procedures can readily be carried out, often by technicians, in the laboratory where the hazardous materials are used.

Bulk quantities of hazardous materials and solutions in various solvents can also often be degraded using the procedures described in this book. Methods for cleaning up spills are frequently indicated, as are solvents for wipe tests to ensure complete surface decontamination. A listing of hazardous compounds, indexed by name, molecular formula and CAS Registry number provides ready access to the information.

The safe handling and disposal of hazardous chemicals is an essential requirement for working with these substances. We hope that this book will contribute to encouraging the use of tested and sound practices.

GEORGE LUNN

ERIC B. SANSONE

Frederick, Maryland
January, 1990

v

CAMROSE LUTHERAN COLLEGE
LIBRARY

Acknowledgments

The primary impetus for our undertaking the kind of work that led to this book was provided by the Division of Safety, National Institutes of Health (NIH), under the leadership first of Dr. W.E. Barkley and later Dr. R.W. McKinney. Dr. M. Castegnaro, of the International Agency for Research on Cancer, has organized collaborative studies with the support of the Division of Safety, NIH in which we have taken part. These studies have contributed materially to this work.

This research was sponsored by the National Cancer Institute under contract No. NO1-CO-74102 with Program Resources, Inc.

We wish to thank Bruce Tobias, Dr. Louise Hellwig, and Dr. Steve Miller for reading the manuscript and making many helpful suggestions.

Contents

DESTRUCTION OF
HAZARDOUS CHEMICALS
IN THE LABORATORY

INTRODUCTION

Biological agents can be completely inactivated by treating them with formaldehyde, ethylene oxide, or moist heat, and radioactive materials will decay with the passage of sufficient time, but there are no general destruction techniques which are applicable to chemical agents. The availability of destruction techniques for hazardous chemical agents would be particularly helpful because of the dangers associated with their handling and disposal. In addition, being able to destroy or inactivate the hazardous materials where they are used is advantageous because the user should be familiar with the hazards of these materials and the precautions required in their handling.

Here we present summaries of destruction procedures for a variety of hazardous chemicals. Most of the procedures have been validated, many by international collaborative testing. We have drawn on information available in the literature[1-11] and on our own published and unpublished work.

ABOUT THIS BOOK

This book is a collection of techniques for destroying a variety of hazardous chemicals. It is intended for those whose knowledge of the chemistry of the compounds covered is rather sophisticated; that is, for those who are aware not only of the obvious dangers, such as the toxic effects of the compounds themselves and of some of the reagents used in the methods, but also of the

1

potential hazards represented, for example, by the possible formation of diazoalkanes when *N*-nitrosamides are treated with base. If you are not thoroughly familiar with the potential hazards and the chemistry of the materials to be destroyed and the reagents to be used, do not proceed.

The destruction methods are organized in what we believe to be rational categories and they are listed in the Table of Contents. It is quite likely, however, that others would have categorized these methods differently, so we have provided three indexes. We have assembled many synonyms of the compounds covered into a Name Index. In each case, the page number given is the first page of the section in which the destruction of that compound is discussed. In some cases, the compound itself may not have been studied; it may have been referred to in the Related Compounds section. Since it is not possible to cite every synonym and every variation in spelling, we have also provided a CAS Registry Number Index and a Molecular Formula Index. With these aids one should be able to find the appropriate destruction method for the compound in question. As a further aid, recognizing the fact that frequently a destruction method is sought only after an accident occurs, we have added an appendix which lists the solvents recommended for use with wipes that are used to sample the area where a spill has occurred in order to determine whether the cleanup has been complete.

One of the difficulties in preparing a book such as this is deciding what should be included and what should be excluded from the text. We have tried to make the method descriptions and the supporting references complete, but at the same time not include unnecessary details. We also tried to eliminate ambiguity wherever possible, going so far as to repeat almost verbatim certain procedures for some compounds rather than noting a minor change and referring to another section and risk a wrong page number or a misinterpretation. Some general safety precautions are given below. These are not repeated for each group of compounds; in some cases, unusual hazards are noted. For many of the destruction procedures we use the word "discard" in connection with the final reaction mixture. This *always* means "discard in compliance with all applicable regulations."

Although we have included all the validated destruction procedures known to us, we realize that there may be other procedures in the literature. Thus, we would be pleased to hear from readers who have any information or suggestions. This work is continuing and we would also be pleased to hear from readers who have suggestions for future work or other comments.

PROPERTIES OF A DESTRUCTION TECHNIQUE

We have already indicated the advantages of destroying hazardous chemicals at the place where they were generated. It is also useful to consider the desirable properties of a destruction technique.

- Destruction of the hazardous chemical should be complete.
- A substantially complete material accountance should be available, with the detectable products being innocuous materials. (This is often difficult to accomplish. In the absence of a complete material accountance, an assessment of the mutagenic activity of the reaction mixture may provide useful information concerning the potential biological hazards associated with the decomposition products.)
- The effectiveness of the technique should be easy to verify analytically.
- The equipment and reagents required should be readily available, inexpensive, and easy and safe to use. The reagents should have no shelf-life limitations.
- The destruction technique should require no elaborate operations (such as distillation or extraction) that might be difficult to contain; it must be easy to perform reliably and should require little time.
- The method should be applicable to the real world; that is, it should be capable of destroying the compound itself, solutions in various solvents, and spills.

These properties characterize an ideal destruction technique. Most techniques cannot meet all of these criteria, but they represent a goal toward which one should strive.

CONTENTS OF A MONOGRAPH

Each monograph usually contains the following information:

- An introduction describes the various properties of the compound or class of compounds being considered.
- The principles of destruction section details, in general terms, the chemistry of the destruction procedures, the products, and the efficiency of destruction.

- The destruction procedures section may be subdivided into procedures for bulk quantities, solutions in water, organic solvents, and so on.
- The analytical procedures section describes one or more procedures that may be used to test the final reaction mixtures to ensure that the compound has been completely degraded. The techniques usually involve packed column gas chromatography (GC) or reverse phase high-pressure liquid chromatography (HPLC) but colorimetric procedures and thin-layer chromatography (TLC) are also used in some cases.
- The mutagenicity assays section describes the data available on the mutagenic activity of the starting materials, possible degradation products, and final reaction mixtures. The data were generally obtained from the plate incorporation technique of the *Salmonella*/mammalian microsome mutagenicity assay (see below).
- The related compounds section describes other compounds to which the destruction procedures should be applicable. The destruction procedures have not usually been validated for these materials, however; they should be fully investigated before adopting them.
- References identify the sources of the information given in the monograph.

MUTAGENICITY ASSAYS

The residues produced by the destruction methods were tested for mutagenicity. Unless otherwise specified, the reaction mixtures from the destruction procedures and some of the starting materials and products were tested for mutagenicity using the plate incorporation technique of the *Salmonella*/mammalian microsome assay essentially as recommended by Ames et al.[12] with the modifications of Andrews et al.[13] Some or all of the tester strains TA98, TA100, TA1530, TA1535, TA1537, and TA1538 of *Salmonella typhimurium* were used with and without S9 rat liver microsomal activation. The reaction mixtures were neutralized before testing. In general, basic reaction mixtures were neutralized by adding acetic acid. Acidic reaction mixtures were neutralized by adding solid sodium bicarbonate. Reaction mixtures containing potassium permanganate were decolorized with sodium ascorbate before neutralization. One hundred microliters of solution (corresponding to varying amounts of undegraded material) was used per plate. Pure compounds were generally tested at a level of 1 mg/

plate in either dimethyl sulfoxide (DMSO) or aqueous solution. One hundred microliters of these solutions were added to each plate. The criterion for significant mutagenicity was set at more than twice the level of the control value. The control value was the average of the cells only and cells plus solvents runs. Unless otherwise specified, residues did not exhibit mutagenic activity. The absence of mutagenic activity in the residual solutions, however, does not necessarily imply that they are nontoxic or have no other adverse biological or environmental effects.

SPILLS

The initial step in dealing with a spill should be the removal of as much of the spill as possible by using a HEPA (high efficiency particulate air) filter equipped vacuum cleaner for solids and absorbents for liquids or solutions. The residue should be decontaminated as described in the monographs.

Whereas solutions or bulk quantities may be treated with heterogeneous [for example, nickel-aluminum (Ni-Al) alloy reduction] or homogeneous methods [for example, potassium permanganate/sulfuric acid ($KMnO_4$/ H_2SO_4) oxidation], decontamination of glassware, surfaces, and equipment and the treatment of spills is best accomplished with homogenous methods. These methods allow the reagent, which is in solution, to contact all parts of the surface to be decontaminated. At the end of the cleanup it is frequently useful to rub the surface with a wipe moistened with a suitable solvent and analyze the wipe for the spilled compound. A list of suitable solvents is given in the Appendix.

APPLICABILITY OF PROCEDURES

Methods that successfully degrade some compounds may not affect other compounds of the same class or other classes of compounds. For example, oxidation with $KMnO_4$ in H_2SO_4 solution has been successfully applied to the destruction of several classes of compounds such as aromatic amines[8] and polycyclic aromatic hydrocarbons.[4] This method gave satisfactory results with some of the antineoplastic agents but not with others, including most of the N-nitrosourea drugs.[9] Sodium hypochlorite treatment, often recommended as a general destruction technique, failed to give satisfactory results with doxorubicin and daunorubicin[9] and polycyclic aromatic hydrocarbons,[4] whereas it did work for aflatoxins.[2] Nickel-aluminum alloy in dilute base worked well for N-nitrosamines[3] but was unsatisfactory for the destruction of polycyclic aromatic hydrocarbons.[4]

Chromic acid is an attractive oxidizing agent and has been used successfully to degrade many compounds, but the spent chromium compounds are potentially carcinogenic. They are also environmentally hazardous and may not be discharged into the sewer. For this reason, we have not recommended the use of chromic acid for degrading any of the compounds we have covered. Potassium permanganate/sulfuric acid degradation appears to be as efficient and has fewer hazards.

SAFETY CONSIDERATIONS

A first step in minimizing risks associated with hazardous chemicals is to prepare a set of guidelines regulating such work. Many organizations have produced such guidelines and many texts have been written on the subject.[14-31] Such documents will provide many useful suggestions when preparing guidelines for any laboratory situation. It is important that the guidelines "fit" the management and administrative structure of the institution and that any particular work requirements be taken into account.

To ensure the safety of those working with hazardous materials of any kind, policies, responsibility, and authority must be clearly defined. The responsibilities of the laboratory director, the supervisor, the employee, and the safety committee should be clearly spelled out.

It is important that potentially hazardous materials are handled only by those workers who have received the appropriate training. For that reason glassware and equipment should be decontaminated in the laboratory before they are transferred to any central washing system.

Obviously, it is important to consider the waste disposal aspects of one's work before the work begins. Experiments should always be designed to use the minimum quantities of potentially hazardous materials, and plans should be made in advance to minimize the wastes generated by any experimentation. Although we concentrate here on laboratory methods for destroying or decontaminating hazardous chemicals, it is valuable to briefly discuss some other approaches to handling chemical wastes. Regardless of the disposal approach selected, only completely decontaminated wastes producing no adverse biological effects should be discarded. Procedures for disposing of hazardous chemicals must comply with all applicable regulations. It is obviously undesirable to deliberately dispose of hazardous chemicals through the sewage system or by evaporation into the atmosphere, unless one has solid evidence that their subsequent degradation is extremely rapid, irreversible, complete, and produces safe degradation products.

It is impossible to provide a concise summary of safety practices for handling hazardous chemicals in the laboratory. For a complete discussion the reader is advised to consult readily available references.[14-31] Each institution and facility should tailor its program to meet its needs. It is important that the safety program include procedures for working with chemicals, biological materials, compressed gases, high voltage power supplies, radioisotopes, and so on.

The following descriptions are designed to give a sufficiently complete guide to the destruction methods available in order to allow one to implement them successfully. The user may wish to consult the sources cited in order to determine the exact reaction conditions, limitations, and hazards that we have not been able to list because of space limitations. In some cases more than one procedure is listed. In these instances all the procedures should be regarded as equally valid unless restrictions on applicability are noted. In the course of collaborative testing we have occasionally found that the efficacy of the same technique varies between laboratories and may also depend on the batch of reagents being used. Thus, we strongly recommend that these methods be periodically validated to ensure that the chemicals are actually being destroyed. These methods have been tested on a limited number of compounds. The efficiency of the destruction techniques must be confirmed when they are applied to a new compound.

The details of analytical techniques are also included. It should be noted that even if 99.5% of a compound is destroyed, the remaining amount may still pose a considerable hazard, particularly if the original reaction was performed on a large scale. The efficiency of degradation is generally indicated by giving the limit of detection, for example, <0.5% of the original compound remained. This means that **none** of the original compound could be detected in the final reaction mixture. However, because of the limitations of the analytical techniques used, it is possible that traces of the original compound, which were below the limit of detection, remained. If this is the case, to use the example given above, the quantity that remained was less than 0.5% of the original amount.

In some cases we found that injecting unneutralized reaction mixtures onto the hot GC column caused degradation of the material for which we were analyzing. Thus it might be that degradation was incomplete but the appropriate peak was not observed in the chromatogram because the compound was degraded on the GC column. Spiking experiments can be used to determine if this is a problem. In a spiking experiment a small amount of the original compound is added to the final reaction mixture and this spiked mixture is analyzed. If an appropriate peak is observed, compound degradation on the GC column is not a problem. If an appropriate peak is not observed, it may be necessary to neutralize the reaction mixture before analysis and/or use a different GC column. Similar problems may be encountered when using HPLC because of the formation of salts or the influence of the sample solvent; again, spiking experiments should be employed. We have indicated in the monographs some instances where problems such as these were encountered (see, for example, Halogenated Compounds) but spiking experiments should be used routinely to test the efficacy of the analytical techniques.

The reactions described were generally performed on the scale specified. If the scale is greatly increased unforeseen hazards may be introduced, particularly with respect to the production of large amounts of heat, which may not be apparent in a small scale reaction. Extra care should therefore be exercised when these reactions are performed on a large scale.

In addition to the potential hazards posed by the compounds themselves, many of the reagents used in degradation procedures are hazardous. All reactions should be carried out in a properly functioning chemical fume hood, which is vented to the outside. Laminar flow cabinets or other recirculating hoods with or without filters are not appropriate. The performance of the hood should be checked by qualified personnel at regular intervals. Hoods should be equipped with an alarm that sounds if the airflow drops below a preset value.

Dissolving concentrated H_2SO_4 in H_2O is a very exothermic process and appropriate protective clothing, including eye protection, should be worn. Concentrated H_2SO_4 should **always** be added to H_2O and **never** the other way around (otherwise splashing of hot concentrated H_2SO_4 may occur). To prepare sulfuric acid solutions the appropriate quantity of concentrated H_2SO_4 is slowly and cautiously added to ~ 500 mL of H_2O, which is stirred in a 1-L flask. When addition is complete water is added to bring the volume up to 1 L and the mixture is allowed to cool to room temperature before use. To prepare a 1 M H_2SO_4 solution use 53 mL of concentrated H_2SO_4 and to prepare a 3 M H_2SO_4 solution use 160 mL of concentrated H_2SO_4.

Appropriate protective clothing should be worn. This includes, but is not limited to, eye protection (safety glasses or face shield), lab coat, and gloves. Rubber gloves generally allow the passage of organic liquids and solutions in organic solvents; they should not be allowed to routinely come into contact with them. Protective clothing should be regarded as the last line of defense and should be changed immediately if it becomes contaminated.

Wastes should be segregated into solid, aqueous, nonchlorinated organic, and chlorinated organic material and disposed of in accordance with local regulations.

In the introductions to the monographs we did not try to give an exhaustive listing of the toxicity data [for example, LD_{50} (the dose which is lethal to 50% of the animals tested) or TLV (threshold limit values) data] or other hazards associated with the compounds under consideration. Instead, we attempted to give some indication of the main hazards associated with each compound or class of compounds. Extensive listings of all the *known* hazards associated with these compounds can be found elsewhere.[14-16]

All organic compounds discussed in this book should be regarded as flammable and all volatile compounds should be regarded as having the capacity of forming explosive mixtures in confined spaces. In many cases the toxic properties of many of these compounds have simply not been adequately investigated. Prudence dictates that, unless there is good reason for believing otherwise, all of the compounds discussed in this book should be regarded as volatile, highly toxic, flammable, human carcinogens, and should be handled with great care.

Other hazards are introduced by the reagents needed to perform the destruction procedures. One example is the use of nickel-aluminum (Ni-Al) alloy that reacts with base to produce hydrogen, a flammable gas that forms explosive mixtures with air. Providing the reactions are done in a fume hood this should not be a problem. It has been found that this reaction frequently exhibits an induction period.[32] There is an initial temperature rise when the Ni-Al alloy is first added but the temperature soon declines to ambient levels. Typically, after ~ 3 h, a much larger temperature rise occurs and the reaction mixture has frequently been observed to boil at this stage. For this reason the reaction should be carried out in a flask that is certainly no more than half-full. In some cases we have observed that considerable foaming occurs and that an even larger flask is required. These instances are mentioned in the monographs (see, for example, Antineoplastic Alkylating Agents). We have found it convenient to perform these reactions in a round-bottom flask fitted with an air condenser. The reaction also produces finely divided nickel, which is potentially pyrophoric. This does not appear to be a problem, however, as long as it is allowed to dry on a metal tray away from flammable solvents for 24 h before being discarded.

Acids and bases are corrosive and should be prepared and used carefully. As noted above, the dilution of concentrated H_2SO_4 is a very exothermic process, which can result in splattering if carried out incorrectly.

A number of potential hazards have been identified. We have made no attempt to provide comprehensive guidelines for safe work, however, and it is essential that workers follow a code of good practice.

REFERENCES

1. National Research Council, Committee on Hazardous Substances in the Laboratory. *Prudent Practices for Disposal of Chemicals from Laboratories;* National Academy Press: Washington, DC, 1983.

2. Castegnaro, M.; Hunt, D.C.; Sansone, E.B.; Schuller, P.L.; Siriwardana, M.G.; Telling, G.M.; van Egmond, H.P.; Walker, E.A., Eds. *Laboratory Decontamination and Destruction of Aflatoxins B_1, B_2, G_1, and G_2 in Laboratory Wastes;* International Agency for Research on Cancer: Lyon, 1980 (IARC Scientific Publications No. 37).

3. Castegnaro, M.; Eisenbrand, G.; Ellen, G.; Keefer, L.; Klein, D.; Sansone, E. B.; Spincer, D.; Telling, G.; Webb, K., Eds. *Laboratory Decontamination and Destruction of Carcinogens in Laboratory Wastes: Some N-Nitrosamines*; International Agency for Research on Cancer: Lyon, 1982 (IARC Scientific Publications No. 43).

4. Castegnaro, M.; Grimmer,G.; Hutzinger,O.; Karcher,W.; Kunte, H.; Lafontaine,M.; Sansone, E. B.; Telling, G.; Tucker, S.P., Eds. *Laboratory Decontamination and Destruction of Carcinogens in Laboratory Wastes: Some Polycyclic Aromatic Hydrocarbons*; International Agency for Research on Cancer: Lyon, 1983 (IARC Scientific Publications No. 49).

5. Castegnaro, M.; Ellen, G.; Lafontaine, M.; van der Plas, H.C.; Sansone, E. B.; Tucker, S.P., Eds. *Laboratory Decontamination and Destruction of Carcinogens in Laboratory Wastes: Some Hydrazines*; International Agency for Research on Cancer: Lyon, 1983 (IARC Scientific Publications No. 54).

6. Castegnaro, M.; Benard, M.; van Broekhoven, L. W.; Fine, D.; Massey, R.; Sansone, E.B.; Smith, P.L.R.; Spiegelhalder, B.; Stacchini, A.; Telling, G.; Vallon, J.J., Eds. *Laboratory Decontamination and Destruction of Carcinogens in Laboratory Wastes: Some N-Nitrosamides*; International Agency for Research on Cancer: Lyon, 1983 (IARC Scientific Publications No. 55).

7. Castegnaro, M.; Alvarez, M.; Iovu, M.; Sansone, E. B.; Telling, G.M.; Williams, D.T., Eds. *Laboratory Decontamination and Destruction of Carcinogens in Laboratory Wastes:*

Some Haloethers; International Agency for Research on Cancer: Lyon, 1984 (IARC Scientific Publications No. 61).

8. Castegnaro, M.; Barek, J.; Dennis, J.; Ellen, G.; Klibanov, M.; Lafontaine, M.; Mitchum, R.; van Roosmalen, P.; Sansone, E.B.; Sternson, L.A.; Vahl, M., Eds. *Laboratory Decontamination and Destruction of Carcinogens in Laboratory Wastes: Some Aromatic Amines and 4-Nitrobiphenyl*; International Agency for Research on Cancer: Lyon, 1985 (IARC Scientific Publications No. 64).

9. Castegnaro, M.; Adams, J.; Armour, M-. A.; Barek, J.; Benvenuto, J.; Confalonieri, C.; Goff, U.; Ludeman, S.; Reed, D.; Sansone, E. B.; Telling, G., Eds. *Laboratory Decontamination and Destruction of Carcinogens in Laboratory Wastes: Some Antineoplastic Agents*; International Agency for Research on Cancer: Lyon, 1985 (IARC Scientific Publications No. 73).

10. Armour, M-.A.; Browne, L.M.; Weir, G.L., Eds. *Hazardous Chemicals. Information and Disposal Guide*, 3rd ed.; University of Alberta: Edmonton, Alberta, 1987.

11. Armour, M-.A.; Browne, L.M.; McKenzie, P.A.; Renecker, D.M.; Bacovsky,R.A., Eds. *Potentially Carcinogenic Chemicals, Information and Disposal Guide*; University of Alberta: Edmonton, Alberta, 1986.

12. Ames, B.N.; McCann, J.; Yamasaki, E. Methods for detecting carcinogens and mutagens with the *Salmonella*/mammalian-microsome mutagenicity test. *Mutat. Res.* **1975**, *31*, 347–364.

13. Andrews, A.W.; Thibault, L.H.; Lijinsky, W. The relationship between carcinogenicity and mutagenicity of some polynuclear hydrocarbons. *Mutat. Res.* **1978**, *51*, 311–318.

14. Sax, N.I; Lewis, R.J., Sr. *Dangerous Properties of Industrial Materials*, 7th ed.; Van Nostrand-Reinhold: New York, 1989.

15. Bretherick, L. *Handbook of Reactive Chemical Hazards*, 3rd ed.; Butterworths: London, 1985.

16. Bretherick, L., Ed. *Hazards in the Chemical laboratory*, 4th ed.; Royal Society of Chemistry: London, 1986.

17. National Research Council, Committee on Hazardous Substances in the Laboratory. *Prudent Practices for Handling Hazardous Chemicals in Laboratories;* National Academy Press: Washington, DC, 1981.

18. Manufacturing Chemists Association. *Guide for Safety in the Chemical Laboratory*, 2nd ed.; Van Nostrand-Reinhold: New York, 1972.

19. Montesano, R.; Bartsch, H.; Boyland, E.; Della Porta, G.; Fishbein, L.; Griesemer, R.A.; Swan, A.B.; Tomatis, L., Eds. *Handling Chemical Carcinogens in the Laboratory: Problems of Safety*; International Agency for Research on Cancer: Lyon, 1979 (IARC Scientific Publications No. 33).

20. Castegnaro, M.; Sansone, E.B. *Chemical Carcinogens*; Springer-Verlag: New York, 1986.

21. Steere, N.V., Ed. *Handbook of Laboratory Safety*, 2nd ed.; CRC Press: Boca Raton, FL, 1971.

22. Young, J.A., Ed. *Improving Safety in the Chemical Laboratory: A Practical Guide*; Wiley: New York, 1987.

23. American Chemical Society, Committee on Chemical Safety. *Safety in Academic Chemistry Laboratories*, 4th ed.; American Chemical Society: Washington, DC, 1985.

24. Pal, S.B., Ed. *Handbook of Laboratory Health and Safety Measures*; Kluwer Academic Publishers: Hingham, MA, 1985.

25. Freeman, N.T.; Whitehead, J. *Introduction to Safety in the Chemical Laboratory*; Academic Press: New York, 1982.

26. LaDou, J., Ed. *Introduction to Occupational Health and Safety;* National Safety Council: Chicago, 1986.

27. Miller, B.M., Ed. *Laboratory Safety: Principles and Practices*; American Society for Microbiology: Washington, DC, 1986.

28. Fuscaldo, A.A.; Erlick, B.J.; Hindman, B., Eds. *Laboratory Safety: Theory and Practice*; Academic Press: New York, 1980.

29. Rosenlund, S.J. *The Chemical Laboratory: Its Design and Operation: A Practical Guide for Planners of Industrial, Medical, or Educational Facilities;* Noyes Publishers: Park Ridge, NJ, 1987.

30. Lees, R.; Smith, A.F., Eds. *Design, Construction, and Refurbishment of Laboratories*; Ellis Horwood: Chichester, England, 1984.

31. DiBerardinis, L.J.; Baum, J.; First, M.W.; Gatwood, G.T.; Groden, E.; Seth, A.K. *Guidelines for Laboratory Design: Health and Safety Considerations*; Wiley: New York, 1987.

32. Lunn, G. Reduction of heterocycles with nickel-aluminum alloy. *J. Org. Chem.* **1987**, *52*, 1043–1046.

ACID HALIDES AND ANHYDRIDES

CAUTION! Refer to safety considerations section on page 6 before starting any of these procedures.

Acid halides such as acetyl chloride [$CH_3C(O)Cl$][1] and benzoyl chloride [$PhC(O)Cl$];[2] sulfonyl chlorides such as benzenesulfonyl chloride [$PhSO_2Cl$][3] and p-toluenesulfonyl chloride [$p\text{-}CH_3C_6H_4SO_2Cl$];[4] and anhydrides such as acetic anhydride [$(CH_3C(O))_2O$],[5] are widely used in organic chemistry. All of these compounds are irritants and corrosive, especially acetyl chloride,[6] acetic anhydride,[7] and benzoyl chloride,[8] which are highly irritating substances. Benzenesulfonyl chloride may explode on storage.[9] They all react readily, and sometimes violently with H_2O, alcohols, dimethyl sulfoxide, and amines. Under controlled conditions these compounds are readily hydrolyzed to the corresponding acids.

Destruction Procedures[10]

To degrade 0.5 mol of the compound stir a sodium hydroxide (NaOH) solution (2.5 M, 600 mL) in a 1-L flask and add a few milliliters of the

13

compound. If the compound dissolves and heat is generated, add the rest of the compound at such a rate that the reaction remains under control. If the reaction is slow (for example, with p-toluenesulfonyl chloride), heat the mixture to $\sim$ 90°C (for example, with a steam bath) and, when the compound has dissolved, add the rest of the compound dropwise. When a clear solution is obtained, allow it to cool. Neutralize the final, cooled, reaction mixture and discard it.

Related Compounds

This procedure should be generally applicable to acid halides, sulfonyl halides, and acid anhydrides.

References

1. Other names are ethanoyl chloride or acetic acid chloride.
2. Other names are benzenecarbonyl chloride, benzoic acid chloride, or α-chlorobenzaldehyde.
3. Other names are benzene sulfone chloride or benzenesulfonic acid chloride.
4. Other names are 4-methylbenzenesulfonyl chloride, tosyl chloride, or toluenesulfonic acid chloride.
5. Other names are acetic oxide, acetyl oxide, ethanoic anhydrate, or acetyl ether.
6. Sax, N.I; Lewis, R.J., Sr. *Dangerous Properties of Industrial Materials*, 7th ed.; Van Nostrand-Reinhold: New York, 1989; p. 46.
7. Reference 6, pp 20–21.
8. Reference 6, pp 403–404.
9. Reference 6, p. 372.
10. National Research Council, Committee on Hazardous Substances in the Laboratory. *Prudent Practices for Disposal of Chemicals from Laboratories;* National Academy Press: Washington, DC, 1983; p. 67.

AFLATOXINS

CAUTION! Refer to safety considerations section on page 6 before starting any of these procedures.

Aflatoxins are fungal metabolites produced by *Aspergillus parasiticus* and *A. flavus*. In hot humid areas peanuts, beans, and corn may be contaminated with aflatoxins. A variety of aflatoxins are known and they are all high-melting ($> 180°C$) crystalline solids. The most commonly encountered aflatoxins are B_1, B_2, G_1, G_2, and M_1 (which is the major metabolite of aflatoxin B_1 in milk). Other aflatoxins are known. They are all chemically very similar[1] and the structure of aflatoxin B_1 is shown.

Aflatoxins are carcinogenic in humans and laboratory animals.[2] They are also acutely poisonous by ingestion.[3] They are used in the laboratory in cancer research and they may also be found as analytical standards in laboratories doing surveillance of foodstuffs. Solid aflatoxins may become electrostatically charged and cling to glassware or protective clothing.

Aflatoxins may be degraded using ammonia (NH_3), potassium permanganate in sulfuric acid ($KMnO_4$ in H_2SO_4), or 5.25% sodium hypochlorite solution followed by the addition of acetone. The acetone is required to destroy any 2,3-dichloroaflatoxin B_1 which may have been formed by the action of the sodium hypochlorite.[4] In addition, animal carcasses may be decontaminated by burying them in quicklime (calcium oxide).[5]

Destruction Procedures[5]

Destruction of Stock Quantities

A. Add sufficient methanol ($\sim$ 1 mL or more if required) to solubilize the aflatoxins and wet the glassware, then add 2 mL of 5.25% sodium hypochlorite solution (see below for assay procedure) for each microgram (μg) of aflatoxin. Allow this to stand overnight, then add an equal volume of H_2O and add a volume of acetone equal to 5% of the total diluted volume. After 30 min check for completeness of destruction and discard it.

B. Add sufficient water so that the aflatoxins are dissolved and their concentration does not exceed 2 μg/mL. Then for each 100 mL of this solution **cautiously** add 10 mL of concentrated H_2SO_4 with stirring (**exothermic reaction!**). Add 16 g of $KMnO_4$ per liter of the resulting solution. The purple color should remain for at least 3 h. If it does not, add more $KMnO_4$. Leave it to react for a further 3 h, then decolorize it with ascorbic acid, neutralize it, test it for completeness of destruction, and discard it.

Destruction of Aflatoxins in Solution

A. Evaporate solutions in organic solvents to dryness (add an equal volume of dichloromethane to dimethyl sulfoxide (DMSO) solutions before evaporation), then solubilize the residual aflatoxins in a little methanol ($\sim$ 1 mL). Treat aqueous solutions as they are. For each microgram of aflatoxin add 2 mL of 5.25% sodium hypochlorite solution (see below for assay procedure). Allow it to stand overnight, then add an equal volume of H_2O and a volume of acetone equal to 5% of the total diluted volume. After 30 min check for completeness of destruction and discard it.

B. Evaporate solutions in organic solvents to dryness (add an equal volume of dichloromethane to DMSO solutions before evaporation), then dissolve the residual aflatoxins in H_2O ($\sim$ 10 mL for each 20 µg of aflatoxins; more if required). Treat aqueous solutions as they are. For each 100 mL of this solution **cautiously** add 10 mL of concentrated H_2SO_4 with stirring (**exothermic reaction!**). Add 16 g of $KMnO_4$ per liter of the resulting solution. The purple color should remain for at least 3 h. If it does not, add more $KMnO_4$. Leave it to react for a further 3 h, then decolorize it with ascorbic acid, neutralize it, test it for completeness of destruction, and discard it.

Destruction of Aflatoxins in Oil

Add 2 mL of a 5.25% sodium hypochlorite solution (see below for assay procedure) for each microgram of aflatoxin, shake the mixture on a mechanical shaker for at least 2 h, add an equal volume of H_2O for each volume of sodium hypochlorite used, then add a volume of acetone equal to 5% of the total diluted volume. After 30 min check for completeness of destruction and discard it.

Decontamination of Equipment and TLC Plates

First rinse equipment with a little methanol to solubilize the aflatoxins. Immerse equipment, TLC plates, protective clothing, and absorbent paper in a 1:1 mixture of 5.25% sodium hypochlorite solution (see below for assay procedure) and H_2O for at least 2 h, then add an amount of acetone equal to 5% of the total volume, allow the mixture to react for at least 30 min, and discard it.

Treatment of Spills

First remove as much of the spill as possible by HEPA vacuuming (not sweeping), then rinse the area with a little methanol to solubilize the aflatoxins. Take up the rinse with absorbent paper. Immerse the absorbent paper in a 1:1 mixture of a 5.25% sodium hypochlorite solution (see below for assay procedure) and H_2O for at least 2 h, then add an amount of acetone equal to 5% of the total volume, allow the mixture to react for at least 30 min, and discard it. Wash the surface from which the spill has been removed with a 5.25% sodium hypochlorite solution and leave it for 10 min before adding a 5% aqueous solution of acetone.

Destruction of Aflatoxins in Animal Litter

Spread the litter on a metal tray to a maximum depth of ~ 5 cm, then sprinkle it with a 5% NH_3 solution (30–40 mL per 25 g of litter). Autoclave the tray for 20 min at 128-130°C, then discard the litter. **Do not** preevacuate the autoclave as this would remove the NH_3.

Destruction of Aflatoxins in Animal Carcasses

Bury carcasses in quicklime and cover to a depth of ~ 1 cm.

Analytical Procedures

Extract 200 mL of decontaminated waste solution three times with 50 mL portions of chloroform and combine the extracts. Concentrate the extracts to ~ 3 mL on a rotary evaporator and add this solution to a graduated tube. Wash the flask twice with 2 mL portions of chloroform and add these washes to the tube. Concentrate the contents of the tube at ~ 60°C to 0.5 mL under a stream of nitrogen. Spot a TLC plate with 10 μL of this solution and with 5 μL of a 0.2 mg/L standard solution of aflatoxins and develop with a mixture of chloroform:acetone (9:1) in subdued light. Determine the presence or absence of aflatoxins by visualizing under ultraviolet (UV) light (365 nm). (The TLC plates used were Kieselgel 60 Merck.) More cleanup of the sample may be required, prior to TLC, if the sample is highly colored or if the aflatoxins were initially dissolved in oil.[5] Other analytical techniques have been reviewed.[6,7]

Mutagenicity Assays

Aflatoxins B_1 and G_1 have been shown to be mutagenic in *Salmonella typhimurium* and other species[8] but specific studies of possible mutagenic products from these degradation reactions have not been carried out.

Related Compounds

The above techniques were investigated for aflatoxins B_1, B_2, G_1, and G_2 but they should also be applicable to other aflatoxins.

Assay of Sodium Hypochlorite Solution

Sodium hypochlorite solutions tend to deteriorate with time, so they should be periodically checked for the amount of active chlorine they contain. Pipette 10 mL of the sodium hypochlorite solution into a 100-mL volumetric flask and fill it to the mark with distilled water. Pipette 10 mL of this solution into a conical flask containing 50 mL of distilled H_2O, 1 g of potassium iodide, and 12.5 mL of 2 M acetic acid. Titrate this solution against a 0.1 N sodium thiosulfate solution using starch as an indicator. Each 1 mL of the sodium thiosulfate solution corresponds to 3.545 mg of active chlorine. The sodium hypochlorite solution used in these degradation reactions should contain 25–30 g of active chlorine/L.

References

1. The systematic names are aflatoxin B_1: 2,3,6aα,9aα-tetrahydro-4-methoxycyclopenta[c]-furo[3′,2′:4,5]furo[2,3-h][1]benzopyran-1,11-dione, aflatoxin B_2: 2,3,6aα,8,9,9aα-hexahydro-4-methoxycyclopenta[c]furo[3′,2′:4,5]furo[2,3-h][1]benzopyran-1,11-dione, aflatoxin G_1 : 3,4,7aα,10aα-tetrahydro-5-methoxy-1H,12H-furo[3′,2′:4,5]furo[2,3-h]pyrano[3,4-c][1]benzopyran-1,12-dione, aflatoxin G_2: 3,4,7aα,9,10,10aα-hexahydro-5-methoxy-1H,12H-furo[3′,2′:4,5]furo[2,3-h]pyrano[3,4-c][1]benzopyran-1,12-dione, and aflatoxin M_1: 2,3,6a,9a-tetrahydro-9a-hydroxy-4-methoxycyclopenta[c]furo[3′,2′:4,5]furo[2,3-h][1]benzopyran-1,11-dione (4-hydroxyaflatoxin B_1).

2. International Agency for Research on Cancer. *IARC Monographs on the Evaluation of the Carcinogenic Risk of Chemicals to Humans, Supplement No. 7, Overall Evaluations of Carcinogenicity: An Updating of* IARC Monographs *Volumes 1 to 42*; International Agency for Research on Cancer: Lyon, 1987; pp 83–87.

3. Sax, N.I; Lewis, R.J., Sr. *Dangerous Properties of Industrial Materials*, 7th ed.; Van Nostrand-Reinhold: New York, 1989; pp 92–95.

4. Castegnaro, M.; Friesen, M.; Michelon, J.; Walker,E.A. Problems related to the use of sodium hypochlorite in the detoxification of aflatoxin B_1. *Am. Ind. Hyg. Assoc. J.* **1981**, *42*, 398–401.

5. Castegnaro, M.; Hunt, D.C.; Sansone, E.B.; Schuller, P.L.; Siriwardana, M.G.; Telling, G.M.; van Egmond, H.P.; Walker, E.A., Eds. *Laboratory Decontamination and Destruction of Aflatoxins B₁, B₂, G₁, and G₂ in Laboratory Wastes;* International Agency for Research on Cancer: Lyon, 1980 (IARC Scientific Publications No. 37).

6. Schuller, P.L.; Horwitz, W.; Stoloff, L. A review of sampling plans and collaboratively studied methods of analysis for aflatoxins. *J. Assoc. Off. Anal. Chem.* **1976**, *59*, 1315–1343.

7. Pohland, A.E.; Thorpe, C.W.; Neshein, S. Newer developments in mycotoxin methodology. *Pure Appl.Chem.* **1979**, *52*, 213–223.

8. Wong, J.J.; Hsieh, D.P.H. Mutagenicity of aflatoxins related to their metabolism and carcinogenic potential. *Proc. Natl. Acad. Sci.* **1976**, *73*, 2241–2244.

ALKALI METALS

CAUTION! Refer to safety considerations section on page 6 before starting any of these procedures.

The alkali metals sodium (Na),[1] potassium (K),[2] and lithium (Li)[3] react violently with water or even moist air to generate hydrogen, which can then be ignited by the heat of the reaction.[2] They are corrosive to the skin and incompatible with many organic and inorganic compounds. Potassium may oxidize on storage and oxidized metal may explode violently when handled or cut.[2] These metals are used in organic synthesis and as drying agents. They require special fire extinguishing procedures.

Principle of Destruction

The metals are allowed to react with an alcohol in a slow and controlled fashion to generate the metal alkoxide and hydrogen. The hydrogen is vented into the fume hood and the metal alkoxide is subsequently hydrolyzed with water to give the metal hydroxide and alcohol.

Destruction Procedures

Caution! These procedures present a high fire hazard and should be conducted in a properly functioning chemical fume hood away from flammable solvents. The presence of a nonflammable board or cloth for smothering the reaction, as well as an appropriate fire extinguisher, may be advisable. If possible, do the reaction in batches to minimize the risk.

Sodium and Lithium

Add the sodium or lithium to cold ethanol at such a rate that the reaction does not become violent. If the reaction mixture becomes viscous and the rate of reaction slows, add more ethanol. When all the sodium has been added stir the reaction mixture until all reaction ceases, then examine carefully for the presence of unreacted metal. If none is found, dilute the mixture with water, neutralize, and discard it.

Potassium

Potassium is the most treacherous of the alkali metals and fires during its destruction are not infrequent. Precautions for its safe handling have been described.[4,5]

Add the potassium to *t*-butyl alcohol[4] at a rate so that the reaction does not become violent. If the reaction mixture becomes viscous and the rate of reaction slows, add more *t*-butyl alcohol. When all the potassium has been added, stir the reaction mixture until all reaction ceases, then examine carefully for the presence of unreacted metal. If none is found, dilute the mixture with water, neutralize, and discard. *t*-Amyl alcohol may also be used.[5] Whichever alcohol is used it is important to use an anhydrous grade. If necessary, the alcohol should be dried before use. Powdered 3 Å molecular sieve has been recommended.[6]

References

1. Sax, N.I; Lewis, R.J., Sr. *Dangerous Properties of Industrial Materials*, 7th ed.; Van Nostrand-Reinhold: New York, 1989; p. 3039.
2. Reference 1, pp 2841–2842.
3. Reference 1, p. 2125.

4. Johnson, W.S.; Schneider, W.P. β-Carbethoxy-γ,γ-diphenylvinylacetic acid. In *Organic Syntheses*; Rabjohn, N., Ed.; Wiley: New York, 1963; Coll. Vol. 4, pp 132–135.

5. Fieser, L.F.; Fieser, M. *Reagents for Organic Synthesis*; Wiley: New York, 1967; Vol. 1, pp 905–906.

6. Burfield, D.R.; Smithers, R.H. Desiccant efficiency in solvent and reagent drying. 7. Alcohols. *J. Org. Chem.* **1983**, *48*, 2420–2422.

ANTINEOPLASTIC ALKYLATING AGENTS

> **CAUTION!** Refer to safety considerations section on page 6 before starting any of these procedures.

The drugs considered in this section are all antineoplastic agents and they all have an N-CH$_2$CH$_2$Cl functionality in common. They are all crystalline solids and are generally moderately soluble in alcohols. Because they are basic they are generally soluble in acid and those compounds having a carboxylic acid group are soluble in base. The water solubility varies.

The compounds considered are:

Mechlorethamine[1]
  [mp of hydrochloride 108–110°C]

(ClCH$_2$CH$_2$)$_2$NCH$_3$.HCl
Hydrochloride soluble in H$_2$O

Melphalan[2]
  [mp 182–183°C]

(I)
Almost insoluble in H$_2$O

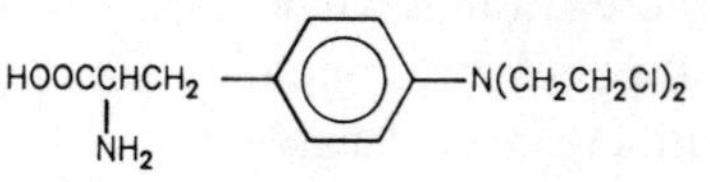

HOOCCHCH$_2$ —◯— N(CH$_2$CH$_2$Cl)$_2$
    |
    NH$_2$

(I)

Chlorambucil[3] **(II)**
[mp 64–66°C]

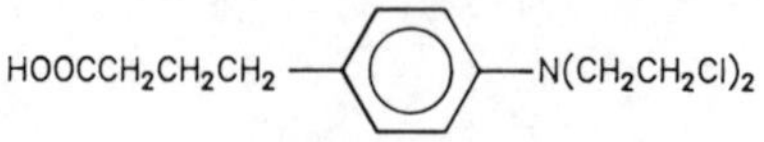

HOOCCH2CH2CH2
N(CH2CH2Cl)2

(II)

Cyclophosphamide[4] **(III)**
[mp 41–45°C (monohydrate)] Soluble in H_2O (40 mg/mL)

Ifosfamide[5] **(IV)**
[mp 39–41°C] Soluble in H_2O (1 in 10)

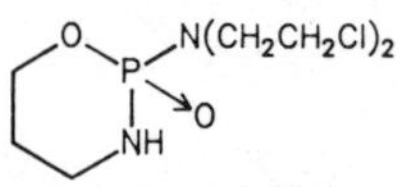

O
P
N(CH2CH2Cl)2
O
NH

(III)

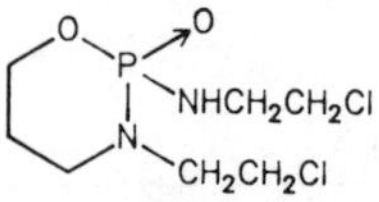

O
P
O
NHCH2CH2Cl
N
CH2CH2Cl

(IV)

Uracil mustard[6] **(V)**
[mp 206°C (decomposes)] Sparingly soluble in H_2O

Spirohydantoin mustard[7] **(VI)**
[mp 127–129°C][8]

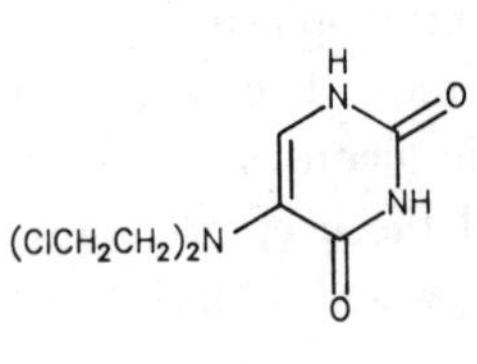

H
N
O
NH
(ClCH2CH2)2N
O

(V)

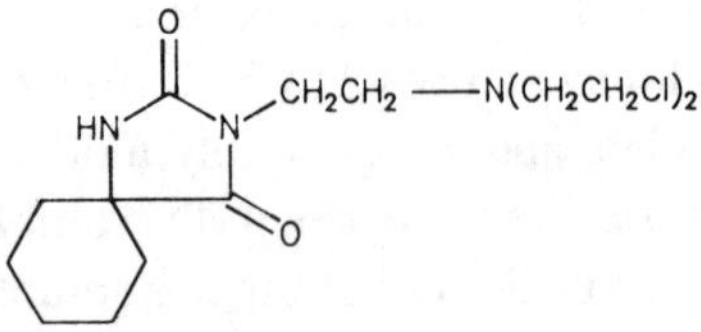

O
HN
N
CH2CH2
N(CH2CH2Cl)2
O

(VI)

Mechlorethamine,[9–11] melphalan,[12–14] chlorambucil,[15–17] cyclophosphamide,[18–21] ifosfamide,[22] and uracil mustard[23,24] are carcinogenic in experimental animals; chlorambucil,[16,17] cyclophosphamide,[19–21] and melphalan[12–14] are human carcinogens; there is limited evidence that mechlorethamine[11] is a human carcinogen. All of these compounds are mutagenic.[25] These compounds are used as antineoplastic drugs.

Principles of Destruction

All the drugs were degraded by reduction with nickel-aluminum (Ni-Al) alloy in potassium hydroxide (KOH) solution.[25] The products detected were ethanol from cyclophosphamide and ifosfamide, and ethylmethylamine and diethylmethylamine from mechlorethamine. Mechlorethamine and chlorambucil were also degraded by reaction with saturated sodium bicarbonate solution ($NaHCO_3$). Pharmaceutical preparations were also degraded using these procedures,[25] but cyclophosphamide tablets first had to be refluxed in 1 M hydrochloric acid (HCl) before the Ni-Al reduction. If this step was omitted the destruction was incomplete. Degradation of mechlorethamine by sodium thiosulfate and $NaHCO_3$ has been recommended,[26] but when the reaction was carried out for the recommended time (45 min) mutagenic products were observed.[25] If the reaction was allowed to proceed for 18 h, however, no mutagenic activity was seen.[25] The International Agency for Research on Cancer (IARC) has recommended alkaline hydrolysis for the degradation of cyclophosphamide and ifosfamide and acid hydrolysis for the degradation of cyclophosphamide,[27] but we found[8] that these reactions gave incomplete degradation and mutagenic products. Cyclophosphamide in urine was degraded using alkaline potassium permanganate ($KMnO_4$) followed by addition of sodium thiosulfate.[28] Melphalan was degraded by oxidation with $KMnO_4$ in basic solution.[29] In all cases destruction was > 99.8% except for pharmaceutical preparations of mechlorethamine (> 90% degradation) and cyclophosphamide in urine (degradation efficiency not reported).[28]

Destruction Procedures

Destruction of Bulk Quantities of Melphalan, Uracil Mustard, and Spirohydantoin Mustard

Dissolve the drug in **methanol** so that the concentration does not exceed 10 mg/mL, then add an equal volume of 2 M KOH solution. For every 20 mL of this basified solution add 1 g of Ni-Al alloy. Add quantities of more than 5 g in portions to prevent the reaction from becoming too violent. Some foaming may occur, so the reaction should be done in a vessel at least five times larger than the final volume. Stir the mixture overnight, then filter through a pad of Celite. Test the filtrate for completeness of destruction, neutralize, and discard it. Allow the spent nickel, which is filtered off, to dry on a metal tray away from flammable solvents for 24 h, then discard it with the solid waste.

Destruction of Bulk Quantities of Melphalan

Dissolve 20 mg of melphalan in 20 mL of a 2 M NaOH solution and add 0.2 g of $KMnO_4$. Stir for 1 h, decolorize with sodium bisulfite, neutralize, check for completeness of destruction, and discard it.

Destruction of Bulk Quantities of Mechlorethamine, Chlorambucil, Cyclophosphamide, and Ifosfamide

Dissolve the drug in **water** so that the concentration does not exceed 10 mg/ mL, then add an equal volume of 2 M KOH solution. For every 20 mL of this basified solution add 1 g of Ni-Al alloy. Add quantities of more than 5 g in portions to prevent the reaction from becoming too violent. Some foaming may occur, so the reaction should be done in a vessel at least five times larger than the final volume. Stir the mixture overnight, then filter through a pad of Celite. Test the filtrate for completeness of destruction, neutralize, and discard it. Allow the spent nickel, which is filtered off, to dry on a metal tray away from flammable solvents for 24 h, then discard it with the solid waste.

Destruction of Pharmaceutical Preparations of Mechlorethamine, Melphalan, Chlorambucil, and Cyclophosphamide

Melphalan. Dissolve the pharmaceutical preparation (100 mg) in the supplied diluent (10 mL), as directed, and add an equal volume of 2 M KOH solution.

Cyclophosphamide. The pharmaceutical preparation consists of 100 mg of drug in 10 mL of saline solution. Add an equal volume of 2 M KOH solution.

Chlorambucil. Dissolve each 2-mg tablet in 10 mL of 1 M KOH solution.

For every 20 mL of basified solution add 1 g of Ni-Al alloy. Add quantities of more than 5 g in portions to prevent the reaction from becoming too violent. Some foaming may occur, so the reaction should be done in a vessel at least five times larger than the final volume. Stir the mixture overnight, then filter through a pad of Celite. Test the filtrate for completeness of destruction, neutralize, and discard it. Allow the spent nickel, which is filtered off, to dry on a metal tray away from flammable solvents for 24 h, then discard it with the solid waste.

Destruction of Mechlorethamine and Chlorambucil

Take up bulk quantities in water so that the concentration does not exceed 10 mg/mL. If necessary, dilute pharmaceutical preparations so that their concentrations do not exceed 10 mg/mL. The chlorambucil may not be completely soluble in the H_2O but it will dissolve when the base is added. Add five volumes of a saturated $NaHCO_3$ solution for each volume of aqueous solution and allow the mixture to stand overnight, check for completeness of destruction, and discard it. Prepare a saturated $NaHCO_3$ solution by mixing $NaHCO_3$ and H_2O in a container. Shake the container occasionally. If solid persists, the solution is saturated, if not, add more $NaHCO_3$.

Destruction of Cyclophosphamide Tablets

For each 50-mg tablet add 10 mL of 1 M HCl and reflux the mixture for 1 h, cool, and place in a vessel whose volume is at least 10 times the final solution volume. Add an equal volume of 2 M KOH solution and stir. For every 20 mL of this basified solution, add 1 g of Ni-Al alloy. Add quantities of > 5 g in portions to prevent the reaction from becoming too violent. Considerable foaming will occur but the reaction should stay in the flask. Stir the mixture overnight, then filter through a pad of Celite. Test the filtrate for completeness of destruction, neutralize, and discard it. Allow the spent nickel, which is filtered off, to dry on a metal tray away from flammable solvents for 24 h, then discard it with the solid waste.

Destruction of Cyclophosphamide in Urine[28]

For each 20 mL of urine add 0.5 mL of 5 M KOH solution followed by 1.2 g of $KMnO_4$. After 2 h add sodium bisulfite until the color of the $KMnO_4$ disappears, then add 1 mL of 5 M KOH solution and 0.66 g of sodium thiosulfate. After 20 min neutralize this mixture by the addition of acid, test for completeness of destruction, and discard it.

Analytical Procedures

Analysis was by HPLC using a 250 × 4.6-mm i.d. column of Microsorb C8. The injection volume was 20 μL and the mobile phase flowed at 1 mL/min.

Ultraviolet (UV) detection, at 254 nm unless otherwise stated, was used. For melphalan, chlorambucil, uracil mustard, and spirohydantoin mustard the mobile phase was a mixture of methanol and 20 mM potassium phosphate, monobasic (KH_2PO_4) buffer: melphalan (58:42); chlorambucil (65:35); uracil mustard (40:60) (UV 200 nm); and spirohydantoin mustard (65:35). It was frequently advantageous to add a little KH_2PO_4 buffer to an aliquot of the neutralized reaction mixture and to centrifuge before analysis. This removed salts that could clog the chromatograph. For cyclophosphamide and ifosfomide the mobile phase was acetonitrile : 20 mM KH_2PO_4 (25:75) and the UV detector was set at 190 nm. Although direct injection could be employed for cyclophosphamide, a massive early eluting peak made detection difficult. It was found advantageous to add solid sodium chloride (NaCl) to 0.5 mL of the reaction mixture and stir. If necessary, more NaCl was added until solid persisted. Acetonitrile (0.25 mL) was then added and the mixture was stirred for 5 min. The acetonitrile layer, which now contained any traces of the drug, was analyzed and a much cleaner chromatogram resulted. On our equipment these mobile phase combinations were found to give reasonable retention times (8–16 min).

Mechlorethamine was determined using a colorimetric procedure. Thus 100 μL of the reaction mixture was mixed with 1 mL of a solution of 2 mL of glacial acetic acid in 98 mL of 2-methoxyethanol and 1 mL of a 5% (w/v) solution of 4-(4-nitrobenzyl)pyridine was added. This mixture was heated at 100°C for 10 min, then cooled in ice for 5 min. Piperidine (0.5 mL) was then added, the mixture centrifuged, and the absorbance determined at 560 nm against an appropriate blank using 10-mm cuvettes. Under these conditions the limit of detection was ~ 50 μg/mL.

Gas chromatography using a 1.8 m × 2-mm i.d. glass column packed with 28% Pennwalt 223 + 4% KOH on 80/100 Gas Chrom R was used to determine the products of these reactions. The injection temperature was 200°C and the flame ionization detector operated at 300°C. The oven temperature was 60°C and the approximate retention times were ethanol (2.7 min), ethylmethylamine (1.5 min), and diethylmethylamine (4.9 min).

Mutagenicity Assays[25]

The mutagenicity assays were carried out as described on page 4 using tester strains TA98, TA100, TA1530, and TA1535. The final reaction mixtures from the Ni-Al alloy reductions were tested at a level corresponding to 0.25 mg (0.125 mg for cyclophosphamide tablets) of undegraded material per plate. To avoid toxicity problems it was generally necessary to mix the neutralized reaction mixtures with an equal volume of pH 7 buffer before testing. The reaction mixtures from $NaHCO_3$ degradation were tested without using buffer and the degradation products from ~ 0.17 mg of drug were applied to each plate. None of the reaction mixtures were found to be mutagenic. All of the drugs were tested at a level of 0.5 mg/plate and were found to be mutagenic. None of the products identified were found to be mutagenic (1 mg/plate). After decontamination with alkaline $KMnO_4$ and sodium thiosulfate, urine which contained cyclophosphamide was not mutagenic to TA98, TA100, UTH8414, and UTH8413.[28] Residues obtained from the alkaline $KMnO_4$ oxidation of melphalan were not mutagenic to TA98, TA100, and TA1535.[29]

Related Compounds

The Ni-Al alloy technique described above should be applicable to compounds of the general form $RR'N-CH_2CH_2Cl$ but the procedure should be thoroughly validated.

References

1. Other names are *N,N*-bis(2-chloroethyl)methylamine, chloramin, di(2-chloroethyl)-methylamine, 2,2'-dichlorodiethyl-*N*-methylamine, 2,2'-dichloro-*N*-methyldiethylamine, *N*-methylbis(2-chloroethyl)amine, *N*-methyl-2,2'-dichlorodiethylamine, methyldi(2-chloroethyl)amine, *N*-methyl-lost, mustine, nitrogen mustard, 2-chloro-*N*-(2-chloroethyl)-*N*-methylethanamine, chlormethine, MBA, and HN2. The compound is generally supplied as the hydrochloride.

2. Other names are 4-[bis(2-chloroethyl)amino]phenylalanine, *p*-di(2-chloroethyl)amino-phenylalanine, phenylalanine mustard, sarcolysine, L-PAM, melfalan, Alkeran, Sarcoclorin, alanine nitrogen mustard, medphalan, merphalan, and phenylalanine nitrogen mustard as well as numerous variants based on the D, L, or DL forms of the phenylalanine.

3. Other names are 4-[bis(2-chloroethyl)amino]benzenebutanoic acid, γ-[*p*-bis(2-chloroethyl)aminophenyl]butyric acid, 4-(*p*-[bis(2-chloroethyl)amino]phenyl)butyric acid, chloraminophene, chloroambucil, chlorobutine, *N,N*-di-2-chloroethyl-γ-*p*-aminophenylbutyric acid, *p*-(*N,N*-di-2-chloroethyl)aminophenylbutyric acid, γ-[*p*-di(2-chloroethyl)amino-phenyl]butyric acid, Amboclorin, Leukeran, and phenylbutyric acid nitrogen mustard.

4. Other names are bis(2-chloroethyl)phosphoramide cyclic propanolamide ester, bis(2-chloroethyl)phosphamide cyclic propanolamide ester, cyclophosphoramide, 1-bis(2-chloroethyl)amino-1-oxo-2-aza-5-oxaphosphoridine, 2-[bis(2-chloroethyl)amino]-2*H*-1,3,2-oxazaphosphorine 2-oxide, 2-[bis(2-chloroethyl)amino]tetrahydro-(2*H*)-1,3,2-oxazaphospho-

rine 2-oxide, *N,N*-bis(2-chloroethyl)-*N'*-(3-hydroxypropyl)phosphorodiamidic acid intramol ester, *N,N*-bis(2-chloroethyl)-*N',O*-propylenephosphoric acid ester diamide, *N,N*-bis(2-chloroethyl)tetrahydro-2*H*-1,3,2-oxazaphosphorin-2-amine 2-oxide (Chemical Abstracts name), *N,N*-bis(2-chloroethyl)-*N',O*-trimethylenephosphoric acid ester diamide, *N,N*-di(2-chloroethyl)amino-*N,O*-propylenephosphoric acid ester diamide, cyclophosphane, cytophosphane, Cytoxan, Endoxan, Procytox, and Sendoxan. The compound is generally supplied as the monohydrate.

5. Other names are *N*,3-bis(2-chloroethyl)tetrahydro-2*H*-1,3,2-oxaphosphorin-2-amine 2-oxide, 3-(2-chloroethyl)-2-[(2-chloroethyl)amino]tetrahydro-2*H*-1,3,2-oxaphosphorin-2-oxide, isophosphamide, iphosphamide, isoendoxan, Cyfos, Holoxan, Mitoxana, and Naxamide.

6. Other names are chlorethaminacil, aminouracil mustard, 5-[bis(2-chloroethyl)amino]-2,4(1*H*,3*H*)-pyrimidinedione, 5-[bis(2-chloroethyl)amino]uracil, 5-[di(2-chloroethyl)-amino]uracil, 2,6-dihydroxy-5-bis(2-chloroethyl)aminopyrimidine, demethyldopan, desmethyldopan, and uramustine.

7. Other names are 3-(2-[bis(2-chloroethyl)amino]ethyl)-1,3-diazaspiro[4,5]decane-2,4-dione, and spiromustine.

8. Lunn, G. Unpublished observations.

9. International Agency for Research on Cancer. *IARC Monographs on the Evaluation of the Carcinogenic Risk of Chemicals to Man.* Volume 9, *Some Aziridines,* N, S- *and* O-*Mustards and Selenium;* International Agency for Research on Cancer: Lyon, 1975; pp 193–207.

10. International Agency for Research on Cancer. *IARC Monographs on the Evaluation of the Carcinogenic Risk of Chemicals to Humans, Supplement No. 4, Chemicals, Industrial Processes and Industries Associated with Cancer in Humans. IARC Monographs, Volumes 1 to 29;* International Agency for Research on Cancer: Lyon, 1982; pp 170–172.

11. International Agency for Research on Cancer. *IARC Monographs on the Evaluation of the Carcinogenic Risk of Chemicals to Humans, Supplement No. 7, Overall Evaluations of Carcinogenicity: An Updating of* IARC Monographs *Volumes 1 to 42;* International Agency for Research on Cancer: Lyon, 1987; pp 269–271.

12. Reference 9, pp 167–180.

13. Reference 10, pp 154–155.

14. Reference 11, pp 239–240.

15. Reference 9, pp 125–134.

16. International Agency for Research on Cancer. *IARC Monographs on the Evaluation of the Carcinogenic Risk of Chemicals to Humans.* Volume 26, *Some Antineoplastic and Immunosuppressive Agents;* International Agency for Research on Cancer: Lyon, 1981; pp 115-136.

17. Reference 11, pp 144–145.

18. Reference 9, pp 135–156.

19. Reference 16, pp 165–202.

20. Reference 10, pp 99–100.

21. Reference 11, pp 182–184.

22. Reference 16, pp 237–247.

23. Reference 11, pp 370–371.

24. Reference 9, pp 235–241.

25. Lunn, G.; Sansone, E.B.; Andrews, A.W.; Hellwig, L.C. Degradation and disposal of some antineoplastic drugs. *J. Pharm. Sci.* **1989**, *78*, 652–659.

26. *Physician's Desk Reference*, 43rd ed.; Medical Economics Co.:Oradell, NJ, 1989; p. 1372.

27. Castegnaro, M.; Adams, J.; Armour, M-. A.; Barek, J.; Benvenuto, J.; Confalonieri, C.; Goff, U.; Ludeman, S.; Reed, D.; Sansone, E. B.; Telling, G., Eds. *Laboratory Decontamination and Destruction of Carcinogens in Laboratory Wastes: Some Antineoplastic Agents*; International Agency for Research on Cancer: Lyon, 1985 (IARC Scientific Publications No. 73).

28. Monteith, D.K.; Connor, T.H.; Benvenuto, J.A.; Fairchild, E.J.; Theiss, J.C. Stability and inactivation of mutagenic drugs and their metabolites in the urine of patients administered antineoplastic therapy. *Environ. Mol. Mutagenesis* **1987**, *10*, 341-356.

29. Barek, J.; Castegnaro, M.; Malaveille, C.; Brouet, I.; Zima, J. A method for the efficient degradation of melphalan into nonmutagenic products. *Microchem. J.* **1987**, *36*, 192–197.

AROMATIC AMINES

> **CAUTION!** Refer to safety considerations section on page 6 before starting any of these procedures.

Aromatic amines constitute a group of widely used synthetic organic chemicals. Many have been shown to be carcinogens in experimental animals and a number are thought to be human carcinogens. 4-Aminobiphenyl,[1,2] benzidine,[3-5] and 2-naphthylamine[6,7] are human and animal carcinogens and 3,3'-dichlorobenzidine,[8-10] 3,3'-dimethoxybenzidine,[11,12] di-(4-amino-3-chlorophenyl)methane,[13] 3,3'-dimethylbenzidine,[14] and 2,4-diaminotoluene[15] cause cancer in laboratory animals. Diaminobenzidine may cause cancer in experimental animals.[16] 3,3'-Dichlorobenzidine may cause cancer in humans.[9,10] In addition, benzidine may cause damage to the blood.[17]

In a recent collaborative study organized by the International Agency for Research on Cancer (IARC) on the laboratory destruction of aromatic amines[18] the following aromatic amines were considered: 4-aminobiphenyl (4-ABP),[19] benzidine (Bz; **I**, R = H),[20] 3,3'-dichlorobenzidine (DClB; **I**, R = Cl),[21] 3,3'-dimethoxybenzidine (DMoB; **I**, R = OCH$_3$),[22] 3,3'-dimethyl-

35

benzidine (DMB; **I**, R = CH$_3$),[23] di(4-amino-3-chlorophenyl)methane (MOCA),[24] 1-naphthylamine (1-NAP),[25] 2-naphthylamine (2-NAP),[26] and 2,4-diaminotoluene (TOL).[27] A procedure for the destruction of diamino-benzidine (DAB; **I**, R = NH$_2$)[28] has also been published.[29]

(I)

All of these compounds are crystalline solids and are generally very sparingly soluble in cold water, more soluble in hot water, and very soluble in acid and in organic solvents. Some of these compounds are used in the chemical industry. They are also used in chemical laboratories and in bio-medical research, for example, as stains and analytical reagents.

Note. Unless otherwise stated all of these procedures can be used for all of the amines mentioned above.

Principles of Destruction

Aromatic amines may be oxidized with potassium permanganate in sulfuric acid (KMnO$_4$ in H$_2$SO$_4$).[18] The products of this reaction have not been identified. Aromatic amines may also be removed from solution using horseradish peroxidase in the presence of hydrogen peroxide (H$_2$O$_2$).[18,30] The enzyme catalyzes the oxidation of the aromatic amine to a radical. These radicals diffuse into solution and polymerize. The polymers are in-soluble and fall out of solution. Although the resulting solution is non-mutagenic, the precipitated polymer is mutagenic so this method was only recommended for the treatment of large quantities of aqueous solution containing small amounts of aromatic amines.[18] In all cases destruction was >99%. Recently, we showed that the above procedures may also be ap-plied to diaminobenzidine. We also found that residual amounts of H$_2$O$_2$ in the horseradish peroxidase procedure produced a mutagenic response and that the mutagenicity could be removed by adding ascorbic acid solution (to reduce the H$_2$O$_2$) to the final reaction mixtures.[31] Various procedures involving diazotization followed by decomposition of the diazo compound have been investigated but the results seem to depend on the nature of the aromatic amine.[18] These procedures are not discussed here.

Destruction Procedures

Destruction of Aromatic Amines in Bulk and in Organic Solvents[18]

Evaporate solutions of aromatic amines in organic solvents to dryness using a rotary evaporator. Dissolve the aromatic amines as follows: For each 9 mg of Bz, DClB, DMB, DMoB, DAB, 1-NAP, 2-NAP, and TOL dissolve in 10 mL of 0.1 M hydrochloric acid (HCl); for each 2.5 mg of MOCA dissolve in 10 mL of 1 M H_2SO_4; for each 2 mg of 4-ABP dissolve in 10 mL of glacial acetic acid; for each 2 mg of mixtures of the above amines add 10 mL of glacial acetic acid. Stir these solutions until the aromatic amines have completely dissolved, then for each 10 mL of the solution so formed add 5 mL of 0.2 M $KMnO_4$ solution and 5 mL of 2 M H_2SO_4. Allow the mixture to stand for at least 10 h, then analyze for completeness of destruction. Decolorize the mixture by the addition of ascorbic acid, neutralize, and discard it.

Destruction of Aromatic Amines in Aqueous Solution[18]

Dilute with water, if necessary, so that the concentration of MOCA does not exceed 0.25 mg/mL, the concentration of 4-ABP does not exceed 0.2 mg/mL, and the concentration of the other amines does not exceed 0.9 mg/mL. For each 10 mL of solution add 5 mL of 0.2 M $KMnO_4$ solution and 5 mL of 2 M H_2SO_4 solution. Allow the mixture to stand for at least 10 h, then analyze for completeness of destruction. Decolorize the mixture by the addition of ascorbic acid, neutralize, and discard it.

Destruction of Aromatic Amines in Oil[18]

Extract the oil solution with 0.1 M HCl until all the amines are removed (at least 2 mL of HCl will be required for each micromole of amine). For each 10 mL of HCl solution add 5 mL of 0.2 M $KMnO_4$ solution and 5 mL of 2 M H_2SO_4 solution. Allow the mixture to stand for at least 10 h, then analyze for completeness of destruction. Decolorize the mixture by the addition of ascorbic acid, neutralize, and discard it.

Decontamination of Spills[18]

Remove as much of the spill as possible by the use of absorbents and HEPA vacuuming, then wet the surface with glacial acetic acid until all the amines are dissolved. Add an excess of a mixture of equal volumes of 0.2 M

KMnO$_4$ solution and 2 M H$_2$SO$_4$ to the spill area. Allow the mixture to stand for at least 10 h, then remove it with absorbents and decolorize the surface with a 5% solution of ascorbic acid.

Decontamination of Glassware[18]

Immerse the glassware in a mixture of equal volumes of 0.2 M KMnO$_4$ and 2 M H$_2$SO$_4$. Allow the glassware to stand in the bath for at least 10 h, then remove it and decolorize the surface with a 5% solution of ascorbic acid.

Decontamination of Large Quantities of Solutions Containing Aromatic Amines[18]

Note. This method is recommended for all the amines listed above except 4-ABP and 2-NAP where destruction was found to be incomplete.

Adjust the pH of aqueous solutions to 5–7 using acid or base as appropriate and dilute so that the concentration of aromatic amines does not exceed 100 mg/L. Dilute solutions in methanol, ethanol, dimethyl sulfoxide, or dimethylformamide with sodium acetate solution (1 g/L) so that the concentration of organic solvent does not exceed 20% and the concentration of aromatic amines does not exceed 100 mg/L. For each liter of solution add 3 mL of a 3% solution of H$_2$O$_2$ and 300 units of horseradish peroxidase. Allow the mixture to stand for 3 h, then remove the precipitate by filtration or centrifugation. For each liter of filtrate add 100 mL of a 5% (w/v) ascorbic acid solution. Check the solution for completeness of degradation and discard. The residue is mutagenic and should be treated as hazardous. It has been reported that further oxidation of this residue by KMnO$_4$ in H$_2$SO$_4$ produces nonmutagenic residues.[18] Filter the reaction mixture through a porous glass filter and immerse this filter in a 1:1 mixture of 0.2 M aqueous KMnO$_4$ solution and 2 M H$_2$SO$_4$ solution.[32]

The horseradish peroxidase used was Donor: hydrogen peroxide oxidoreductase; EC 1.11.1.7 (Type II) having a specific activity of 150–200 purpurogallin units/mg obtained from Sigma. An appropriate amount was dissolved in sodium acetate solution (1 g/L), then an aliquot of this solution was used to obtain the requisite number of units.

Analytical Procedures

There are many publications on the analysis of aromatic amines. The following HPLC analysis was recommended by the IARC.[18] A 250 × 4.6-mm reverse phase column was used and the mobile phase was acetonitrile :

methanol : buffer (10:30:20) flowing at 1.5 mL/min. The buffer was 1.5 mM in potassium phosphate, dibasic (K_2HPO_4) and 1.5 mM in potassium phosphate, monobasic (KH_2PO_4). If a variable wavelength UV detector is available the following wavelengths can be used: Bz = 285 nm, DMoB = 305 nm, DMB = 285 nm, DClB = 285 nm, 1-NAP = 240 nm, 2-NAP = 235 nm, 4-ABP = 275 nm, MOCA = 245 nm, and TOL = 235 nm. If only a fixed wavelength detector is available, use 280 nm for the first four; 254 nm for the rest. We have found that diaminobenzidine can be determined using the above buffer : methanol (75:25) flowing at 1 mL/min with the UV detector set at 300 nm.[31]

Mutagenicity Assays[18]

Reaction mixtures obtained when the aromatic amines (except for diaminobenzidine) were degraded with $KMnO_4$ in H_2SO_4 were tested for mutagenicity using *Salmonella typhimurium* strains TA97, TA98, and TA100. No mutagenic activity was seen. Using the same strains, the supernatants from the horseradish peroxidase-H_2O_2 reactions were tested. Again, no mutagenic activity was seen but the solid residues from Bz, DClB, DMoB, and 2-NAP were mutagenic. Reaction mixtures obtained from the degradation of diaminobenzidine were tested using strains TA98, TA100, TA1530, and TA1535 and no mutagenic activity was found.

Related Compounds

The procedures described above should be generally applicable to other aromatic amines but thorough validation is essential in each case as some variability may be observed, particularly for the horseradish peroxidase method.[18]

References

1. International Agency for Research on Cancer. *IARC Monographs on the Evaluation of the Carcinogenic Risk of Chemicals to Humans, Supplement No. 7, Overall Evaluations of Carcinogenicity: An Updating of* IARC Monographs *Volumes 1 to 42*; International Agency for Research on Cancer: Lyon, 1987; pp 91–92.

2. International Agency for Research on Cancer. *IARC Monographs on the Evaluation of Carcinogenic Risk of Chemicals to Man.* Volume 1; International Agency for Research on Cancer: Lyon, 1971; pp 74–79.

3. Reference 1, pp 123–125.

4. Reference 2, pp 80–86.

5. International Agency for Research on Cancer. *IARC Monographs on the Evaluation of the Carcinogenic Risk of Chemicals to Humans*. Volume 29, *Some Industrial Chemicals and Dyestuffs*; International Agency for Research on Cancer: Lyon, 1982; pp 149–183.

6. Reference 1, pp 261–263.

7. International Agency for Research on Cancer. *IARC Monographs on the Evaluation of Carcinogenic Risk of Chemicals to Man*. Volume 4, *Some Aromatic Amines, Hydrazine and Related Substances, N-Nitroso Compounds and Miscellaneous Alkylating Agents*; International Agency for Research on Cancer: Lyon, 1974; pp 97–111.

8. Reference 1, pp 193–194.

9. Reference 7, pp 49–55.

10. Reference 5, pp 239–256.

11. Reference 1, pp 198–199.

12. Reference 7, pp 41–47.

13. Reference 1, pp 246–247.

14. Reference 2, pp 87–91.

15. International Agency for Research on Cancer. *IARC Monographs on the Evaluation of Carcinogenic Risk of Chemicals to Man*. Volume 16, *Some Aromatic Amines and Related Nitro Compounds - Hair Dyes, Colouring Agents and Miscellaneous Industrial Chemicals*; International Agency for Research on Cancer: Lyon, 1978; pp 83–95.

16. Weisburger, E.K.; Russfield, A.B.; Homburger, F.; Weisburger, J.H.; Boger, E.; Van Dongen, C.G.; Chu, K.C. Testing of twenty-one environmental aromatic amines or derivatives for long-term toxicity or carcinogenicity. *J. Environ. Path. Toxicol.* **1978**, *2*, 325–356.

17. Sax, N.I; Lewis, R.J., Sr. *Dangerous Properties of Industrial Materials*, 7th ed.; Van Nostrand-Reinhold: New York, 1989; pp 376–377.

18. Castegnaro, M.; Barek, J.; Dennis, J.; Ellen, G.; Klibanov, M.; Lafontaine, M.; Mitchum, R.; van Roosmalen, P.; Sansone, E.B.; Sternson, L.A.; Vahl, M., Eds. *Laboratory Decontamination and Destruction of Carcinogens in Laboratory Wastes: Some Aromatic Amines and 4-Nitrobiphenyl*; International Agency for Research on Cancer: Lyon, 1985 (IARC Scientific Publications No. 64).

19. Other names are (1,1'-biphenyl)-4-amine, 4-aminodiphenyl, anilinobenzene, xenylamine, 4-biphenylamine, and *p*-phenylaniline.

20. Other names are (1,1'-biphenyl)-4,4'-diamine, 4,4'-biphenyldiamine, 4,4'-diaminobiphenyl, 4,4'-diaminodiphenyl, and *p,p*'-dianiline.

21. Other names are 4,4'-diamino-3,3'-dichlorobiphenyl, 3,3'-dichloro-4,4'-biphenyldiamine, 3,3'-dichlorobiphenyl-4,4'-diamine, DCB, and 3,3'-dichloro-4,4'-diaminobiphenyl.

22. Other names are 3,3'-dimethoxy(1,1'-biphenyl)-4,4'-diamine, 3,3'-dimethoxy-4,4'-diaminobiphenyl, *o*-dianisidine and *o,o*'-dianisidine.

23. Other names are 3,3'-dimethyl(1,1'-biphenyl)-4,4'-diamine, 3,3'-tolidine, bianisidine, 4,4'-bi-*o*-toluidine, 4,4'-diamino-3,3'-dimethylbiphenyl, 4,4'-diamino-3,3'-dimethyldiphenyl, 3,3'-dimethyl-4,4'-biphenyldiamine, 3,3'-dimethyl-4,4'-diphenyldiamine, 3,3'-

dimethylbiphenyl-4,4'-diamine, 3,3'-dimethyldiphenyl-4,4'-diamine, *o*-tolidine, and *o,o'*-tolidine.

24. Other names are 4,4'-methylenebis(2-chlorobenzenamine), 4,4'-diamino-3,3'-dichlorodiphenylmethane, 4,4'-methylene(bis)chloroaniline, methylene-4,4'-bis(*o*-chloroaniline), *p,p*'-methylenebis(α-chloroaniline), 4,4'-methylenebis(*o*-chloroaniline), *p,p*'-methylenebis(*o*-chloroaniline), 4,4'-methylenebis(2-chloroaniline), and DACPM.

25. Other names are 1-naphthalenamine, α-naphthylamine, 1-aminonaphthalene, and naphthalidine.

26. Other names are 2-naphthalenamine, β-naphthylamine, 2-aminonaphthalene, and 6-naphthylamine.

27. Other names are toluene-2,4-diamine, *m*-tolylenediamine, 3-amino-*p*-toluidine, 5-amino-*o*-toluidine, 1,3-diamino-4-methylbenzene, 2,4-diamino-1-methylbenzene, 4-methyl-1,3-benzenediamine, 4-methyl-*m*-phenylenediamine, 2,4-tolamine, *m*-toluenediamine, 2,4-toluenediamine, *m*-toluylenediamine, 2,4-toluylenediamine, 2,4-tolylenediamine, and tolylene-2,4-diamine.

28. Other names are 3,3',4,4'-biphenyltetramine, 3,3',4,4'-diphenyltetramine, and 3,3',4,4'-tetraaminobiphenyl.

29. Barek, J. Destruction of carcinogens in laboratory wastes. II. Destruction of 3,3'-dichlorobenzidine, 3,3'-diaminobenzidine, 1-naphthylamine, 2-naphthylamine, and 2,4-diaminotoluene by permanganate. *Microchem. J.* **1986**, *33*, 97–101.

30. Klibanov, A.M.; Morris, E.D. Horseradish peroxidase for the removal of carcinogenic aromatic amines from water. *Enzyme Microb. Technol.* **1981**, *3*, 119–122.

31. Lunn, G. Unpublished observations.

32. Barek, J.; Pacáková, V.; Štulík, K.; Zima, J. Monitoring of aromatic amines by HPLC with electrochemical detection. Comparison of methods for destruction of carcinogenic aromatic amines in laboratory wastes. *Talanta* **1985**, *32*, 279–283.

AZIDES

CAUTION! Refer to safety considerations section on
page 6 before starting any of these procedures.

Sodium azide (Smite, NaN_3) can decompose explosively on heating and can
form shock sensitive and highly explosive azides when it comes in contact
with heavy metals.[1] For this reason solutions of NaN_3 should **never** be
poured down the sink. Sodium azide is also acutely toxic.[2] A death has
been reported when water preserved with NaN_3 was ingested in the laboratory.[3] NaN_3 is a powerful mutagen. Treatment with acid liberates explosive,
toxic volatile (bp 37°C) hydrazoic acid (HN_3).[4]

Organic azides vary greatly in stability but a number are known to
decompose explosively with shock or heating.[5] They should be handled
very carefully.

Principles of Destruction

Sodium azide can be decomposed by reaction with ceric ammonium nitrate
or by reaction with nitrous acid. The latter reaction generates toxic oxides
of nitrogen.

43

Organic azides (for example, phenyl azide, PhN_3) can be reduced to the corresponding amines with tin in hydrochloric acid (HCl) or with stannous chloride in methanol.

Destruction Procedures

Sodium Azide

A. For each gram of NaN_3, dissolve 9 g of ceric ammonium nitrate in 170 mL of H_2O.[6] Cool the ceric ammonium nitrate solution in an ice bath and stir. Dissolve the NaN_3 in H_2O and add it in portions. When the addition is complete the reaction mixture should still be orange and adding a small amount of ceric ammonium nitrate should cause no visible reaction.[7] Stir the reaction mixture for 1 h, then discard it.

B. Dissolve NaN_3 (5 g) in at least 100 mL of H_2O. Stir the reaction mixture and add 7.5 g of sodium nitrite dissolved in 38 mL of H_2O.[8] Slowly add dilute H_2SO_4 (20%) until the reaction mixture is acidic to litmus. When no more oxides of nitrogen are evolved test the reaction mixture with starch-iodide paper. If the paper turns blue, indicating the presence of excess nitrous acid, discard the reaction mixture. If the paper does not turn blue, add more sodium nitrite.

Note. It is important to add the sodium nitrite, **then** the H_2SO_4. Adding these reagents in the reverse order will generate explosive, volatile, toxic, hydrazoic acid.

Organic Azides

A. Suspend 10.9 g of stannous chloride with stirring in 40 mL of methanol and add 4 g (0.03 mol) of benzyl azide dropwise.[9] When addition is complete stir the reaction mixture at room temperature for 30 min, then dilute with H_2O, neutralize, and discard it.

B. Slowly add 1 g (0.0084 mol) phenyl azide to a stirred mixture of 6 g of granular tin in 100 mL of concentrated HCl.[10] Stir the mixture for 30 min after addition is complete, then pour it into a large quantity of cold H_2O, neutralize, and discard it. Unreacted tin may be recycled or discarded with the solid waste.

Related Compounds

The procedures described above for NaN_3 should not be used for heavy metal azides, many of which are shock sensitive explosives. Professional help should be sought in these cases. The procedures for organic azides should be generally applicable although in all cases the reactions should be thoroughly validated to ensure that the azides are completely destroyed. The products of these reactions are the corresponding amines, which may themselves be hazardous compounds.

References

1. Bretherick, L. *Handbook of Reactive Chemical Hazards*, 3rd ed.; Butterworths: London, 1985; p. 1305.

2. Sax, N.I; Lewis, R.J., Sr. *Dangerous Properties of Industrial Materials*, 7th ed.; Van Nostrand-Reinhold: New York, 1989; p. 3046.

3. *Chas Notes*, Newsletter of the American Chemical Society Division of Chemical Health and Safety, Vol. 6, No. 3, July-Sept, 1988.

4. Reference 2, pp 1898–1899.

5. Reference 2, pp 327, 410, and 2725–2726.

6. Manufacturing Chemists Association. *Laboratory Waste Disposal Manual*; Manufacturing Chemists Association: Washington, DC, 1973; p. 136.

7. B.I.Tobias, Chemical Safety Officer, NCI-Frederick Cancer Research Facility, personal communication, December 1988.

8. National Research Council, Committee on Hazardous Substances in the Laboratory. *Prudent Practices for Disposal of Chemicals from Laboratories;* National Academy Press: Washington, DC, 1983; p. 88.

9. Maiti, S.N.; Singh, M.P.; Micetich, R.G. Facile conversion of azides to amines. *Tetrahedron Lett.* **1986**, *27*, 1423–1424.

10. Armour, M-.A.; Browne, L.M.; Weir, G.L., Eds. *Hazardous Chemicals. Information and Disposal Guide,* 3rd ed.; University of Alberta: Edmonton, Alberta, 1987; p. 291.

AZO AND AZOXY
COMPOUNDS AND
TETRAZENES

CAUTION! Refer to safety considerations section on
page 6 before starting any of these procedures.

Azobenzene [Ph-N═N-Ph, mp 68°C],[1] azoxybenzene [Ph-N═N(O)-Ph,
mp 36°C],[2] and N,N-dimethyl-4-amino-4′-hydroxyazobenzene [**I**, mp 201–
202°C] are solids and azoxymethane [CH_3-N═N(O)-CH_3, bp 98°C] and
tetramethyltetrazene [$(CH_3)_2$N-N═N-N$(CH_3)_2$, bp 130°C] are liquids. Azobenzene and azoxybenzene are insoluble in H_2O and soluble in alcohols
and organic solvents. Azoxymethane is carcinogenic in animals[3] and
azobenzene may be carcinogenic in animals.[4,5] Azoxybenzene is toxic by
ingestion and is a skin and eye irritant;[6] tetramethyltetrazene explodes
when heated to its boiling point.[7] Azoxymethane is used in cancer research
and the other compounds are used in organic synthesis and as chemical
intermediates in the chemical industry.

$$>N-\!\!\!\bigcirc\!\!\!-\!N\!=\!N-\!\!\!\bigcirc\!\!\!-OH$$

(I)

Principles of Destruction

N,N-Dimethyl-4-amino-4'-hydroxyazobenzene can be oxidized by potassium permanganate in sulfuric acid ($KMnO_4$ in H_2SO_4). Destruction is >99.5%; the products have not been identified.[8] The other compounds can be reduced to their parent amines by using nickel-aluminum (Ni-Al) alloy in potassium hydroxide (KOH) solution.[9] Destruction is >99% for tetramethyltetrazene, >99.9% for azobenzene, and >99.5% for the other compounds.[10] When performing these reactions it should be noted that in some cases the products are aromatic amines, which may also be hazardous. For example, the product of the reduction of azobenzene and azoxybenzene is aniline for which limited evidence of carcinogenicity in animals exists.[11]

Destruction Procedures

Destruction of Azobenzene, Azoxybenzene, Azoxymethane, and Tetramethyltetrazene

Dissolve the compound in methanol (azobenzene and azoxybenzene) or H_2O (others) so that the concentration does not exceed 10 mg/mL and add an equal volume of 1 M KOH solution. For each 100 mL of basified solution add 5 g of Ni-Al alloy. Add quantities of more than 5 g in portions over ~ 1 h to avoid excessive frothing. Stir the reaction mixture overnight, then filter through a pad of Celite. Neutralize the filtrate, test for completeness of destruction, and discard it. Place the nickel that is filtered off on a metal tray away from flammable solvents for 24 h, then discard it with the solid waste.

Destruction of N,N-Dimethyl-4-amino-4'-hydroxyazobenzene

Take up 0.24 mg of N,N-dimethyl-4-amino-4'-hydroxyazobenzene in 1 mL of 50% (v/v) acetic acid and add 1 mL of 2 M H_2SO_4 and 1 mL of 0.2 M $KMnO_4$ solution. After 2 h decolorize with oxalic acid, test for completeness of destruction, neutralize and discard it.

Analytical Procedures

For analysis by gas chromatography a 1.8 m × 2-mm i.d. packed column was used with flame ionization detection.[10] For methylamine (retention time 0.9 min at 60°C), dimethylamine (1.3 min at 100°C), azoxymethane (2.6 min at 120°C), and tetramethyltetrazene (3.4 min at 150°C) the packing was 28% Pennwalt 223 + 4% KOH on 80/100 Gas Chrom R and for aniline (2.9 min at 80°C), azobenzene (1.5 min at 180°C), and azoxybenzene (3.2 min at 180°C) the packing was 3% SP 2401-DB on 100/120 Supelcoport. These chromatographic conditions are only a guide; the exact conditions would have to be determined experimentally. N,N-Dimethyl-4-amino-4'-hydroxyazobenzene was determined by differential pulse polarography of the reaction mixture.[8]

Mutagenicity Assays

No data have been reported.

Related Compounds

4-Phenylazophenol, 4-phenylazoaniline, and 4,4'-azoxyanisole have been reduced to their parent amines in moderate to good yield by Ni-Al alloy in KOH solution although the complete disappearance of the starting material has not been established.[9] Nickel-aluminum alloy in KOH solution appears to be a general method for the cleavage of N-N and N-O bonds.[9] Potassium permanganate in H_2SO_4 is a general oxidative procedure and it should be applicable to many azo compounds.[8] In each case full validation should be carried out before using the procedure.

References

1. Other names are azobenzide, azobenzol, azobisbenzene, azodibenzene, azodibenzeneazofume, benzeneazobenzene, and 1,2-diphenyldiazene.

2. Other names are diphenyldiazene 1-oxide, azobenzene oxide, azoxybenzide, and azoxydibenzene.

3. Narisawa, T.; Wong, C.-Q.; Weisburger, J.H. Azoxymethane-induced liver hemangiosarcomas in inbred strain-2 guinea pigs. *J. Natl. Cancer Inst.* **1976**, *56*, 653–654.

4. International Agency for Research on Cancer. *IARC Monographs on the Evaluation of Carcinogenic Risk of Chemicals to Man.* Volume 8, *Some Aromatic Azo Compounds*; International Agency for Research on Cancer: Lyon, 1975, pp 75–81.

5. Sax, N.I; Lewis, R.J., Sr. *Dangerous Properties of Industrial Materials*, 7th ed.; Van Nostrand-Reinhold: New York, 1989; pp 331–332.

6. Reference 5, pp 333–334.

7. Reference 5, p. 3228.

8. Barek, J.; Kelnar, L. Destruction of carcinogens in laboratory wastes. IV. Destruction of *N,N*-dimethyl-4-amino-4′-hydroxyazobenzene by permanganate. *Microchem. J.* **1986**, *33*, 239–242.

9. Lunn, G.; Sansone, E.B.; Keefer, L.K. General cleavage of N-N and N-O bonds using nickel/aluminum alloy. *Synthesis* **1985**, 1104–1108.

10. Lunn, G. Unpublished results.

11. International Agency for Research on Cancer. *IARC Monographs on the Evaluation of the Carcinogenic Risk of Chemicals to Humans, Supplement No. 7, Overall Evaluations of Carcinogenicity: An Updating of IARC Monographs Volumes 1 to 42*; International Agency for Research on Cancer: Lyon, 1987; pp 99–100.

BORON TRIFLUORIDE AND INORGANIC FLUORIDES

> **CAUTION!** Refer to safety considerations section on page 6 before starting any of these procedures.

Soluble inorganic fluorides are highly toxic[1] (for example, the LD_{50} of sodium fluoride is 180 mg/kg[2]). Boron trifluoride (BF_3) is highly toxic[3] and, when supplied as the etherate, it is corrosive and flammable.[4,5]

Take up these compounds in H_2O and add a solution of calcium chloride.[6] Filter the reaction mixture and dispose of the insoluble (solubility 15 mg/L), nontoxic ($LD_{50} > 2.5$ g/kg),[7] calcium fluoride with the solid waste. (Degradation of BF_3 will also produce boric acid.) Remove any organic liquid present and discard the aqueous solution.

References

1. Sax, N.I; Lewis, R.J., Sr. *Dangerous Properties of Industrial Materials*, 7th ed.; Van Nostrand-Reinhold: New York, 1989; p. 1735.

51

2. Smyth, H.F.; Carpenter, C.P.; Weil, C.S.; Pozzani, U.C.; Striegel, J.A.; Nycum, J.S. Range-finding toxicity data: List VII. *Am. Ind. Hyg. Assoc. J.* **1969**, *30*, 470–476.

3. Reference 1, pp 546–547.

4. Reference 1, p. 548.

5. Bretherick, L., Ed. *Hazards in the Chemical laboratory*, 4th ed.; Royal Society of Chemistry: London, 1986; p. 202.

6. National Research Council, Committee on Hazardous Substances in the Laboratory. *Prudent Practices for Disposal of Chemicals from Laboratories;* National Academy Press: Washington, DC, 1983; p. 86.

7. Reference 1, p. 681.

CALCIUM CARBIDE

> **CAUTION!** Refer to safety considerations section on page 6 before starting any of these procedures.

Calcium carbide (acetylenogen, calcium acetylide, CaC_2) is used in the laboratory to generate acetylene. Calcium carbide reacts with small quantities of H_2O to generate acetylene and when this is done in an uncontrolled fashion explosive acetylene-air mixtures may be formed. Decomposition under controlled conditions generates acetylene, which is vented into the fume hood, and calcium chloride solution.

Destruction Procedure[1]

Place CaC_2 (50 g) in a 2-L flask and suspend it by stirring in 600 mL of toluene or cyclohexane. Surround the flask with an ice bath and pass nitrogen into the flask. Exhaust the acetylene that is generated through a plastic tube into the back of the fume hood. Add hydrochloric acid (6 M, 300 mL) dropwise from a dropping funnel over ~ 5 h. Stir the mixture for another hour, then neutralize the aqueous layer, separate the layers, and discard them.

Reference

1. National Research Council, Committee on Hazardous Substances in the Laboratory. *Prudent Practices for Disposal of Chemicals from Laboratories;* National Academy Press: Washington, DC, 1983; p. 96.

CARBAMIC ACID ESTERS

> **CAUTION!** Refer to safety considerations section on page 6 before starting any of these procedures.

Four carbamic acid esters were investigated: methyl carbamate [MC, urethylane, $CH_3OC(O)NH_2$]; ethyl carbamate, which is more commonly called urethane or urethan [UT, $CH_3CH_2OC(O)NH_2$]; N-methylurethane [MUT, $CH_3CH_2OC(O)NHCH_3$]; and N-ethylurethane [EUT, $CH_3CH_2OC(O)NHCH_2CH_3$]. Methyl carbamate (mp 56–58°C) and urethane (mp 48.5–50°C) are solids and N-methylurethane (bp 170°C) and N-ethylurethane (bp 176°C) are liquids. However, liquified methyl carbamate (bp 176–177°C) and urethane (bp 182–184°C) have fairly low boiling points, so all these compounds should be regarded as volatile and they should only be handled inside a chemical fume hood. Urethane,[1] N-methylurethane,[2] and N-ethylurethane[2] are carcinogenic in experimental animals. Evidence for the carcinogenicity of methyl carbamate is inconclusive.[3] Urethane is a teratogen, causes depression of bone marrow and nausea, and can affect the brain and central nervous system.[4] Urethane is used industrially as an intermediate. It also appears to form naturally by

fermentation in some alcoholic beverages. Methyl carbamate is used industrially in the textile and pharmaceutical industries. All of these compounds are soluble in H_2O and ethanol.

Principles of Destruction

These compounds are all hydrolyzed using 5 M sodium hydroxide (NaOH)
solution,[5] although the reaction times vary. The compounds are hydrolyzed
to methanol (MC) or ethanol (UT, MUT, EUT) and either carbamic acid
(MC, UT), N-methylcarbamic acid (MUT), or N-ethylcarbamic acid
(EUT), which then decompose to carbon dioxide and ammonia (MC, UT),
methylamine (MUT), or ethylamine (EUT). In all cases destruction is
>99% and good accountances are obtained for the products.

Destruction Procedures

Destruction of N-Methylurethane, Methyl Carbamate, and Urethane

Add 50 mg of the compound to 10 mL of 5 M NaOH solution and stir at
room temperature for 24 h (MC, UT) or 48 h (MUT). Check the reaction
mixture for completeness of destruction, neutralize, and discard it.
Note. This procedure is **not** suitable for N-ethylurethane.

Destruction of N-Methylurethane, N-Ethylurethane, Methyl Carbamate, and Urethane

Add 50 mg of the compound to 10 mL of 5 M NaOH solution and reflux for
4 h. Cool, check the reaction mixture for completeness of destruction,
neutralize, and discard it.

Analytical Procedures

The carbamic acid esters were determined by gas chromatography using a
1.8 m × 2-mm i.d. packed column filled with 5% Carbowax 20 M on 80/100
Chromosorb W HP.[5] The column was fitted with a precolumn and it was
found helpful to change it regularly. The oven temperature was 120°C

(MUT and EUT) or 140°C (MC and UT), the injection temperature was 200°C, and the flame ionization detector operated at 300°C. Injection of reaction mixtures containing 5 M NaOH solution onto the hot GC column degraded any residual carbamate and gave unreliable results. To get around this, 2 mL of the reaction mixture were acidified, before analysis, with 1 mL of concentrated hydrochloric acid (**Caution!** Exothermic) and this mixture was then neutralized by adding solid sodium bicarbonate. Injection of this solution onto the column gave reliable results. The absence of any carbamate could be confirmed by spiking the reaction mixture with a small amount of a dilute solution of the carbamate in question. The products of the reaction were determined using the same conditions using a column packed with 10% Carbowax 20 M + 2% KOH in 80/100 Chromosorb W AW with an oven temperature of 150°C.

The GC conditions given above are only a guide; the exact conditions would have to be determined experimentally. Using 5 µL injections our detection limits were ~ 30 µg/mL (MC), 7 µg/mL (UT), and 4 µg/mL (MUT and EUT).

Mutagenicity Assays

The mutagenicity assays were carried out as described on page 4 using tester strains TA98, TA100, TA1530, and TA1535. The final reaction mixtures (tested at a level corresponding to 0.5 mg of undegraded material per plate) were not mutagenic. The only pure compound which was mutagenic was MC (tested in dimethyl sulfoxide solution), which was mutagenic to TA98 with activation. Ammonium carbamate, which is related to a possible intermediate in the degradation reactions, was not mutagenic.

Related Compounds

The procedure should be generally applicable to the destruction of carbamic acid esters, but it should be carefully checked to ensure that the compounds are completely degraded. The resistance to hydrolysis appears to increase as the degree of substitution increases. More highly substituted carbamic esters may require prolonged refluxing.

Summary of Reaction Conditions for the Hydrolysis of RNHC(O)OR'

Compound	Conditions	Time (h)	RNH_2 (%)	R'OH (%)	Amount Remaining (%)
MC	Room temperature	24	-	98	<0.61
	Reflux	4	-	75	<0.61
UT	Room temperature	24	-	103	<0.15
	Reflux	4	-	51	<0.16
MUT	Room temperature	48	60	92	<0.075
	Reflux	4	39	69	<0.075
EUT	Reflux	4	43	74	<0.15

References

1. International Agency for Research on Cancer. *IARC Monographs on the Evaluation of Carcinogenic Risk of Chemicals to Man.* Volume 7, *Some Anti-thyroid and Related Substances, Nitrofurans and Industrial Chemicals*; International Agency for Research on Cancer: Lyon, 1974; pp 111–140.

2. Larsen, C.D. Pulmonary-tumor induction with alkylated urethans. *J. Natl. Cancer Inst.* **1948**, *9*, 35–37.

3. International Agency for Research on Cancer. *IARC Monographs on the Evaluation of Carcinogenic Risk of Chemicals to Man.* Volume 12, *Some Carbamates, Thiocarbamates and Carbazides*; International Agency for Research on Cancer: Lyon, 1976; pp 151–159.

4. Sax, N.I; Lewis, R.J., Sr. *Dangerous Properties of Industrial Materials*, 7th ed.; Van Nostrand-Reinhold: New York, 1989; pp 3449–3450.

5. Lunn, G.; Sansone, E. B. Validated methods for degrading hazardous chemicals: Some alkylating agents and other compounds. *J. Chem. Educ.* (in press).

CARBON DISULFIDE

CAUTION! Refer to safety considerations section on page 6 before starting any of these procedures.

Carbon disulfide (carbon bisulfide, dithiocarbonic anhydride, CS_2) is widely used as a solvent. It is highly volatile (bp 46°C), highly toxic, and it readily forms explosive or ignitable mixtures with air.[1,2]

Principle of Destruction

Carbon disulfide is oxidized by hypochlorite to carbon dioxide and sulfuric acid.

Destruction Procedures[2]

A. Stir a 5.25% sodium hypochlorite solution (6.7 L) at room temperature in a 10-L flask and add 30 mL (0.5 mol) of CS_2 dropwise. Regulate the temperature at ~ 20–30°C. When the addition is complete, stir the mixture for 2 h, then discard it. Use fresh sodium hypochlorite solution (see below for assay procedure).

59

CAMROSE LUTHERAN COLLEGE
LIBRARY

B. Stir calcium hypochlorite (550 g) in 2.2 L of H_2O at room temperature in a 5-L flask. Most will soon dissolve. Add CS_2 (30 mL, 0.5 mol)
dropwise. Regulate the temperature at 20–30°C. When the addition is
complete, stir the mixture for 2 h, then discard it.

Assay of Sodium Hypochlorite Solution

Sodium hypochlorite solutjons tend to deteriorate with time so they should
be periodically checked for the amount of active chlorine they contain.
Pipette 10 mL of sodium hypochlorite solution into a 100-mL volumetric
flask, which is filled to the mark with distilled H_2O. Pipette 10 mL of this
solution into a conical flask containing 50 mL of distilled H_2O, 1 g of
potassium iodide, and 12.5 mL of 2 M acetic acid. Titrate this solution
against 0.1 N sodium thiosulfate solution using starch as an indicator. Each
1 mL of the sodium thiosulfate solution corresponds to 3.545 mg of active
chlorine. The sodium hypochlorite solution used in these degradation reactions should contain 25–30 g of active chlorine/L.

References

1. Sax, N.I; Lewis, R.J., Sr. *Dangerous Properties of Industrial Materials*, 7th ed.; Van
   Nostrand-Reinhold: New York, 1989; pp 711–712.
2. National Research Council, Committee on Hazardous Substances in the Laboratory. *Prudent Practices for Disposal of Chemicals from Laboratories;* National Academy Press:
   Washington, DC, 1983; p. 66.

CHLOROMETHYLSILANES

CAUTION! Refer to safety considerations section on page 6 before starting any of these procedures.

The chloromethylsilanes chlorotrimethylsilane (or trimethylsilyl chloride) $((CH_3)_3SiCl)$,[1] dichlorodimethylsilane $((CH_3)_2SiCl_2)$,[2] and methyltrichlorosilane (CH_3SiCl_3)[3] are flammable, volatile, toxic, corrosive liquids used in organic chemistry. They should only be handled in a properly functioning chemical fume hood.

Destruction Procedure[4]

Hydrolyze by cautiously adding to vigorously stirred H_2O in a flask. The reaction produces hydrochloric acid and polymeric silicon-containing material. Remove any insoluble material and discard it with the solid or liquid waste. Neutralize the aqueous layer and discard it.

CHLOROMETHYLSILANES

References

1. Sax, N.I; Lewis, R.J., Sr. *Dangerous Properties of Industrial Materials*, 7th ed.; Van Nostrand-Reinhold: New York, 1989; p. 899.
2. Reference 1, p. 1146.
3. Reference 1, p. 2400.
4. Patnode, W.; Wilcock, D.F. Methylpolysiloxanes. *J. Am. Chem. Soc.* **1946**, *68*, 358–363.

CHROMIUM(VI)

CAUTION! Refer to safety considerations section on page 6 before starting any of these procedures.

Chromium(VI) is an oxidizer and a carcinogen in humans and experimental animals.[1-3] It is widely used in organic synthesis and is one of the hazardous constituents of chromic acid. Because compounds containing Cr(VI) are powerful oxidizers they can react violently with a variety of organic and inorganic compounds.[4] Chromic acid and its salts are poisonous and corrosive to the skin and mucous membranes forming ulcers that are slow to heal.[4]

The compounds for which this procedure has been validated are chromium trioxide [CrO_3, chromium(VI) oxide, chromic anhydride], sodium dichromate ($Na_2Cr_2O_7.2H_2O$, sodium bichromate), potassium dichromate ($K_2Cr_2O_7$, potassium bichromate), ammonium dichromate [$(NH_4)_2Cr_2O_7$, ammonium bichromate], chromic acid [a solution of Cr(VI) in concentrated sulfuric acid (H_2SO_4)], and the commercially available solution Chromerge, both in the concentrated and diluted forms.

63

Principles of Destruction

Chromium(VI) is reduced to Cr(III) (which is not an oxidizer) using sodium metabisulfite and the Cr(III) is precipitated by basification with magnesium hydroxide ($Mg(OH)_2$). The International Agency for Research on Cancer has reported that there is inadequate evidence that Cr(III) is carcinogenic.[3] Sodium or potassium hydroxide give a gelatinous precipitate, which is hard to filter. Precipitates that are easier to filter can be obtained by careful control of the pH,[5] but $Mg(OH)_2$ automatically produces the right pH and the sludgelike precipitate is relatively easy to filter.[6] The clear filtrate is just slightly basic (pH 7.1 - 9.2) and contains no trace of Cr(VI) (<0.25 ppm) and only trace amounts of Cr(III).

Destruction Procedures

Disposal of Bulk Quantities of Chromium(VI)-Containing Compounds (Sodium Dichromate, Potassium Dichromate, Ammonium Dichromate, Chromium Trioxide, and Chromerge Concentrate)

Stir the chromium compound (5 g) in 100 mL of 0.5 M H_2SO_4. When it has completely dissolved add 10 g of sodium metabisulfite. Stir the mixture for 1 h and allow to cool, then check for the presence of Cr(VI). Mix a few drops of the reaction mixture with a few drops of 100 mg/mL potassium iodide (KI) solution. A dark color indicates that Cr(VI) is still present. (If necessary, the color can be made more apparent by adding a drop of starch solution.) If Cr(VI) is still present, add sodium metabisulfite until a negative test is obtained. Add $Mg(OH)_2$ (6 g) to the reaction mixture and stir the mixture for 1 h, then allow it to stand overnight. Decant the mixture into a suction filter apparatus so that the clear liquid is filtered first, then the green precipitate is sucked dry. If the filtrate is yellow, this may indicate the presence of Cr(VI). Check using the KI test described above (acidify first with a little dilute H_2SO_4). If Cr(VI) is present in the filtrate, acidify with H_2SO_4 and repeat the process. The filtrate contains no trace of Cr(VI) and only traces of Cr(III). The sludge does not contain Cr(VI) but does contain large amounts of Cr(III). The sludge is, however, no longer an oxidizer. Dispose of appropriately.

If the reaction is performed on a larger scale than that described above, considerable heat is generated, particularly when the sodium metabisulfite is added, and it may be necessary to let a longer time elapse between stages to allow for cooling.

Disposal of Solutions Containing Chromium(VI) (For Example, New or Used Chromic Acid or Chromerge Solutions)

Carefully add the chromium solution (10 mL), with stirring, to 60 mL of H_2O and stir the mixture for at least 1 h until cool. Add sodium metabisulfite solution (100 mg/mL, 10 mL) and stir the mixture for a few minutes, then check for the presence of Cr(VI). Mix a few drops of the reaction mixture with a few drops of 100 mg/mL KI solution. A dark color indicates that Cr(VI) is still present. (If necessary the color can be made more apparent by adding a drop of starch solution.) If Cr(VI) is still present, add sodium metabisulfite until a negative test is obtained. Add $Mg(OH)_2$ (12 g) to the reaction mixture and stir the mixture for 1 h, then allow it to stand overnight. Decant the mixture into a suction filter apparatus so that the clear liquid is filtered first, then the green precipitate is sucked dry. If the filtrate is yellow, this may indicate the presence of Cr(VI). Check using the KI test described above (acidify first with a little dilute H_2SO_4). If Cr(VI) is present, acidify the filtrate with H_2SO_4 and repeat the process. The filtrate contains no trace of Cr(VI) and only traces of Cr(III). The sludge does not contain Cr(VI) but does contain large amounts of Cr(III). The sludge is, however, no longer an oxidizer. Dispose of appropriately.

If the reaction is performed on a larger scale than that described above, considerable heat is generated, particularly when the solution is initially diluted and when the $Mg(OH)_2$ is added, and it may be necessary to do these procedures more slowly and let a longer time elapse between stages to allow for cooling.

Analytical Procedures

Total chromium can be determined by flame atomic absorption spectroscopy using a hollow cathode lamp at 357.9 nm.

Chromium(VI) may be determined by using a colorimetric procedure.[7] Dissolve sym-diphenylcarbazide (0.20 g) in 100 mL of ethanol to prepare a reagent solution. Add 200 μL of 3 M H_2SO_4 to 3 mL of the reaction mixture and check to make sure that this mixture is acidic. Add 100 μL of the reagent solution and shake the mixture for a few seconds. Let it stand for 10 min, then determine the violet color at 540 nm against a suitable blank. About 0.25 ppm Cr(VI) produces a just visible violet color; high concentrations of Cr(VI) produce a very intense violet color that fades rapidly. If high concentrations of Cr(VI) are present, dilute the sample mixture and repeat the procedure. The response is linear to 4 ppm. The method is quite insensitive to Cr(III), but the method can be adapted to measuring total chromium by oxidizing all the chromium present to Cr(VI).[7]

Mutagenicity Assays

To test the filtrates mutagenicity assays were carried out as described on page 4 using tester strains TA98, TA100, TA1530, and TA1535. One hundred microliters of filtrate was applied to each plate. No mutagenic activity was found.

Related Compounds

This procedure is specific for Cr(VI) and should not be used for any other heavy metal.

References

1. International Agency for Research on Cancer. *IARC Monographs on the Evaluation of Carcinogenic Risk of Chemicals to Man*. Volume 2, *Some Inorganic and Organometallic Compounds*; International Agency for Research on Cancer: Lyon, 1972; pp 100–125.

2. International Agency for Research on Cancer. *IARC Monographs on the Evaluation of the Carcinogenic Risk of Chemicals to Humans*. Volume 23, *Some Metals and Metallic Compounds*; International Agency for Research on Cancer: Lyon, 1980; pp 205–323.

3. International Agency for Research on Cancer. *IARC Monographs on the Evaluation of the Carcinogenic Risk of Chemicals to Humans, Supplement No. 7, Overall Evaluations of Carcinogenicity: An Updating of* IARC Monographs *Volumes 1 to 42*; International Agency for Research on Cancer: Lyon, 1987; pp 165–168.

4. Sax, N.I; Lewis, R.J., Sr. *Dangerous Properties of Industrial Materials*, 7th ed.; Van Nostrand-Reinhold: New York, 1989; pp 912–913.

5. Armour, M-.A.; Browne, L.M.; Weir, G.L. Tested disposal methods for chemical wastes from academic laboratories. *J. Chem. Educ.* **1985**, *62*, A93-A95.

6. Lunn, G.; Sansone, E.B. A laboratory procedure for the reduction of Chromium(VI) to Chromium(III). *J. Chem. Educ.* **1989**, *66*, 443–445.

7. *Standard Methods for the Examination of Water and Wastewater*, 16th ed.; American Public Health Association: Washington DC, 1985; pp 201–204.

CISPLATIN

CAUTION! Refer to safety considerations section on page 6 before starting any of these procedures.

The degradation of a number of antineoplastic drugs, including cisplatin, was investigated by the International Agency for Research on Cancer (IARC).[1] Cisplatin[2] [cis-Pt(NH$_3$)$_2$Cl$_2$] is a solid (mp 270°C) slightly soluble in H$_2$O but insoluble in most organic solvents, except dimethylformamide (DMF). Cisplatin is mutagenic and carcinogenic in animals[3–5] and it is a teratogen and has effects on the bone marrow and kidney.[6] It is used as an antineoplastic drug.

Principles of Destruction

Cisplatin can be destroyed by reduction to elemental platinum with zinc (~ 99% destruction) or by forming an inactive complex with sodium diethyldithiocarbamate. In the latter case the extent of destruction is unknown but the reaction mixtures formed were not mutagenic. Oxidation of cisplatin with potassium permanganate was found to give mutagenic residues.[1]

67

Destruction Procedures

Destruction of Bulk Quantities of Cisplatin

A. Dissolve the cisplatin in 2 M sulfuric acid (H_2SO_4) solution so that its concentration does not exceed 0.6 mg/mL. Stir the reaction mixture and add 3 g of zinc powder, in portions to avoid frothing, for each 100 mL of solution. Stir the reaction mixture overnight, check for completeness of destruction, neutralize, and discard it.

B. Dissolve the cisplatin in H_2O and for every 100 mg of cisplatin add 30 mL of a 0.68 M solution of sodium diethyldithiocarbamate in 0.1 M sodium hydroxide (NaOH). Add 30 mL of saturated aqueous sodium nitrate (NaNO$_3$) solution (a yellow precipitate may form), check for completeness of destruction, and discard the mixture.

Destruction of Cisplatin in Aqueous Solutions and Injectable Pharmaceutical Preparations of 5% Dextrose or 0.9% Saline

A. Dilute the solution with H_2O so that the concentration of the drug does not exceed 0.6 mg/mL and add concentrated H_2SO_4, with stirring, until a 2 M solution is obtained. After cooling, stir the reaction mixture and add 3 g of zinc powder, in portions to avoid frothing, for each 100 mL of solution. Stir the reaction mixture overnight, check for completeness of destruction, neutralize, and discard it.

B. For every 100 mg of cisplatin add 30 mL of a 0.68 M solution of sodium diethyldithiocarbamate in 0.1 M NaOH. Then add 30 mL of saturated aqueous NaNO$_3$ solution (a yellow precipitate may form), check for completeness of destruction, and discard the mixture.

Destruction of Cisplatin in Organic Solvents Miscible with H_2O

Add at least an equal volume of 4 M H_2SO_4, more if required, so that the concentration of the drug does not exceed 0.6 mg/mL. Stir the reaction mixture and add 3 g of zinc powder, in portions to avoid frothing, for each 100 mL of solution. Stir the reaction mixture overnight, neutralize, check for completeness of destruction, and discard it.

Destruction of Cisplatin in Urine[7]

For each 3 mL of urine containing cisplatin add 1 mL of a 10% solution of sodium diethyldithiocarbamate in 0.1 M NaOH and 2 mL of saturated aqueous $NaNO_3$ solution. After 10 min check for completeness of destruction and discard the mixture.

Decontamination of Glassware Contaminated with Cisplatin

A. Rinse the glassware at least four times with enough H_2O to completely wet the glass. Combine the rinses and dilute with H_2O, if necessary, so that the concentration of cisplatin does not exceed 0.6 mg/mL. Add concentrated H_2SO_4, with stirring, until a 2 M solution is obtained. After cooling, stir the reaction mixture and add 3 g of zinc powder, in portions to avoid frothing, for each 100 mL of solution. Stir the reaction mixture overnight, check for completeness of destruction, neutralize, and discard it.

B. Immerse the glassware in a mixture of equal volumes of sodium diethyldithiocarbamate solution (0.68 M in 0.1 M NaOH) and $NaNO_3$ solution (saturated aqueous). Check for completeness of destruction and discard it.

Decontamination of Spills of Cisplatin

Allow any organic solvent to evaporate and remove as much of the spill as possible by HEPA vacuuming (not sweeping), then rinse the area with H_2O. Take up the rinse with absorbent material and allow the rinse and any absorbent material used to react with a mixture of equal volumes of sodium diethyldithiocarbamate solution (0.68 M in 0.1 M NaOH) and $NaNO_3$ solution (saturated aqueous). Check for completeness of decontamination by using a wipe moistened with H_2O. Analyze the wipe for the presence of the drug.

Analytical Procedures

Cisplatin presents an analytical problem. Solutions can be analyzed by atomic absorption using an acetylene-air flame and a platinum lamp at 260 nm but this only shows the presence or absence of platinum. It does not show if the platinum exists as cisplatin or if it has been complexed in an inactive form. Since cisplatin is highly mutagenic, the absence of significant mutagenicity (defined as a number of revertants that is more than twice the background) is a reasonable test for the absence of cisplatin. A 5 μg/mL aqueous solution of cisplatin produced a significant mutagenic response in TA100 with or without activation. If atomic absorption is not available, mix 9 mL of the solution to be tested with 1 mL of 10% sodium diethyldithiocarbamate in 0.1 M NaOH solution and 1 mL of saturated aqueous $NaNO_3$ solution.[1] Shake the reaction mixture and allow to react for 1 h. Add 1 mL of H_2O-saturated chloroform and shake. Centrifuge this mixture (1200 × g) for 5 min, mix on a vortex mixer, and centrifuge for 10 min more. Analyze the organic layer by HPLC using a 30 cm × 3.6-mm i.d. column of μ Bondapack CN with heptane : isopropanol (82:18) flowing at 2 mL/min as the mobile phase and a UV detector set at 254 nm. This technique will not distinguish between platinum in cisplatin and platinum bound in an inactive form.

Other methods of analyzing for cisplatin have been reviewed.[8] Electrochemical detection,[9,10] differential pulse amperometric detection,[11] and post-column reaction detection[12] have all been recommended for use with HPLC.

Mutagenicity Assays

In the IARC study[1] tester strains TA98, TA100, and TA1535 of *Salmonella typhimurium* were used with and without metabolic activation. The reaction mixtures were not mutagenic. When cisplatin in urine was treated, tester strains TA98, TA100, UTH8413, and UTH8414 were used without metabolic activation. No mutagenic activity was observed.[7]

Related Compounds

These procedure may be applicable to other platinum containing compounds, but any new application should be thoroughly validated both for complete destruction of the compound and for the production of nonmutagenic reaction mixtures.

References

1. Castegnaro, M.; Adams, J.; Armour, M-. A.; Barek, J.; Benvenuto, J.; Confalonieri, C.; Goff, U.; Ludeman, S.; Reed, D.; Sansone, E. B.; Telling, G., Eds. *Laboratory Decontamination and Destruction of Carcinogens in Laboratory Wastes: Some Antineoplastic Agents*; International Agency for Research on Cancer: Lyon, 1985 (IARC Scientific Publications No. 73).

2. Other names are *cis*-platinous diammine dichloride, *cis*-diamminedichloroplatinum, *cis*-dichlorodiammine platinum(II), *cis*-platinum(II) diaminedichloride, *cis*-DDP, CACP, CPDC, DDP, Cisplatyl, Neoplatin, Platinex, Platinol, and Peyrone's chloride.

3. International Agency for Research on Cancer. *IARC Monographs on the Evaluation of the Carcinogenic Risk of Chemicals to Humans, Supplement No. 4, Chemicals, Industrial Processes and Industries Associated with Cancer in Humans. IARC Monographs, Volumes 1 to 29*; International Agency for Research on Cancer: Lyon, 1982; pp 93–94.

4. International Agency for Research on Cancer. *IARC Monographs on the Evaluation of the Carcinogenic Risk of Chemicals to Humans.* Volume 26, *Some Antineoplastic and Immunosuppressive Agents*; International Agency for Research on Cancer: Lyon, 1981; pp 151-164.

5. International Agency for Research on Cancer. *IARC Monographs on the Evaluation of the Carcinogenic Risk of Chemicals to Humans, Supplement No. 7, Overall Evaluations of Carcinogenicity: An Updating of* IARC Monographs *Volumes 1 to 42*; International Agency for Research on Cancer: Lyon, 1987; pp 170–171.

6. Sax, N.I; Lewis, R.J., Sr. *Dangerous Properties of Industrial Materials*, 7th ed.; Van Nostrand-Reinhold: New York, 1989; pp 2808–2809.

7. Monteith, D.K.; Connor, T.H.; Benvenuto, J.A.; Fairchild, E.J.; Theiss, J.C. Stability and inactivation of mutagenic drugs and their metabolites in the urine of patients administered antineoplastic therapy. *Environ. Mol. Mutagenesis* **1987**, *10*, 341-356.

8. Riley, C.M.; Sternson, L.A. Recent advances in the clinical analysis of cisplatin. *Pharm. Int.* **1984**, *5*, 15–19.

9. Parsons, P.J.; LeRoy, A.F. Determination of *cis*-diamminedichloroplatinum(II) in human plasma using ion-pair chromatography with electrochemical detection. *J. Chromatogr.* **1986**, *378*, 395–408.

10. Parsons, P.J.; Morrison, P.F.; LeRoy, A.F. Determination of platinum-containing drugs in human plasma by liquid chromatography with reductive electrochemical detection. *J. Chromatogr.* **1987**, *385*, 323–335.

11. Elferink, F.; van der Vijgh, W.J.F.; Pinedo, H.M. Analysis of antitumour [1,1-bis(aminomethyl)cyclohexane]platinum(II) complexes derived from spiroplatin by high-performance liquid chromatography with differential pulse amperometric detection. *J. Chromatogr.* **1985**, *320*, 379–392.

12. Marsh, K.C.; Sternson, L.A.; Repta, A.J. Post-column reaction detector for platinum(II) antineoplastic agents. *Anal. Chem.* **1984**, *56*, 491–497.

COMPLEX METAL HYDRIDES

> **CAUTION!** Refer to safety considerations section on page 6 before starting any of these procedures.

Complex metal hydrides are widely used in organic synthesis. They are generally air and water sensitive and may be pyrophoric. Vigorous reaction with water may ignite the flammable hydrocarbon solvents generally used with these reagents. On reaction with water, flammable hydrogen is released.[1] Calcium hydride (CaH_2),[2] lithium aluminum hydride ($LiAlH_4$, lithium aluminohydride, lithium aluminum tetrahydride, lithium alanate, aluminum lithium hydride, LAH, Lithal, or lithium tetrahydroaluminate),[3] sodium hydride (NaH),[4] and sodium borohydride ($NaBH_4$ or sodium tetrahydroborate)[5] can react violently with a variety of organic and inorganic compounds. Sodium borohydride is toxic by ingestion.[5] Lithium aluminum hydride may ignite when ground.[3] Fires involving $LiAlH_4$ should be smothered with dry sand.[6]

Principles of Destruction

The hydride is allowed to react slowly with H_2O, an alcohol, or ethyl acetate under controlled conditions. Sodium borohydride is more stable and it is necessary to use acetic acid to cause decomposition.

Destruction Procedures

Lithium Aluminum Hydride (LiAlH₄)

A. Stir the $LiAlH_4$ in a suitable solvent and for each n grams of LAH present, slowly add n mL of H_2O under nitrogen.[7] Use an ice bath. Add n mL of 15% sodium hydroxide solution and 3n mL of H_2O in succession and stir the mixture vigorously for 20 min. Filter the granular precipitate that forms. Separate the organic and aqueous layers of the filtrate and discard them.

B. Stir the $LiAlH_4$ under nitrogen in a suitable solvent using an ice bath and slowly add 95% ethanol, which reacts less vigorously than H_2O.[8] When the reaction is complete, filter the mixture, and separate the organic and aqueous layers of the filtrate and discard them.

C. Stir the $LiAlH_4$ under nitrogen in a suitable solvent using an ice bath and slowly add ethyl acetate, which reacts less vigorously than H_2O and generates no hydrogen.[9] When the reaction is complete, add saturated ammonium chloride solution, filter the mixture, and separate the organic and aqueous layers of the filtrate and discard them.

Sodium Borohydride (NaBH₄)

Sodium borohydride is relatively stable in H_2O and acid is needed for its decomposition.[8] Dissolve the solid compound in H_2O and dilute aqueous solutions with H_2O, if necessary, so that the concentration does not exceed 3%. Add dilute aqueous acetic acid dropwise with stirring under nitrogen. Discard when the reaction has ceased.

Sodium Hydride (NaH)

Sodium hydride is generally supplied as a dispersion in mineral oil. Add a dry hydrocarbon solvent (for example, heptane) so that the hydride concentration does not exceed 5% and stir the mixture under nitrogen. Add excess t-butyl alcohol dropwise followed by cold H_2O.[8] Separate the layers and discard them.

Calcium Hydride (CaH$_2$)

For each gram of hydride add 25 mL of methanol, with stirring, under nitrogen. When the reaction has finished add at least an equal volume of H$_2$O and discard it.[8]

Related Compounds

Complex metal hydrides vary greatly in their reactivity and the application of any of the above methods to another hydride should be carefully, and cautiously, investigated before employing it. Nevertheless, the principle of allowing the hydride to react with a less reactive substrate than H$_2$O in a slow and controlled fashion should be generally applicable.

References

1. Sax, N.I; Lewis, R.J., Sr. *Dangerous Properties of Industrial Materials*, 7th ed.; Van Nostrand-Reinhold: New York, 1989; p. 1899.

2. Reference 1, p. 682.

3. Reference 1, pp 2132–2133.

4. Reference 1, p. 3069.

5. Reference 1, p. 3049.

6. Bretherick, L., Ed. *Hazards in the Chemical laboratory*, 4th ed.; Royal Society of Chemistry: London, 1986; p. 169.

7. Micovic, V.M.; Mihailovic, M.L. The reduction of acid amides with lithium aluminum hydride. *J. Org. Chem.* **1953**, *18*, 1190–1200.

8. National Research Council, Committee on Hazardous Substances in the Laboratory. *Prudent Practices for Disposal of Chemicals from Laboratories;* National Academy Press: Washington, DC, 1983; p. 85.

9. Fieser, L.F.; Fieser, M. *Reagents for Organic Synthesis*; Wiley: New York, 1967; Vol. 1, p. 584.

CYANIDES AND CYANOGEN BROMIDE

> **CAUTION!** Refer to safety considerations section on page 6 before starting any of these procedures.

Inorganic cyanides are acutely toxic compounds, for example, the LD_{50} in the rat is 15 mg/kg for sodium cyanide $(NaCN)$[1] and 10 mg/kg for potassium cyanide.[2] These compounds should be handled with great care and not allowed to come in contact with acid that will generate hydrogen cyanide (HCN), a volatile, highly toxic, flammable gas,[3] which forms explosive mixtures with air. Acute cyanide poisoning is rapidly fatal. Less acute cases may produce symptoms such as headache, dizziness, nausea, and so on.[3,4] Inorganic cyanides have a number of uses in organic synthesis. Cyanogen bromide (bromine cyanide, $CNBr$) is a highly toxic, volatile crystalline solid (mp 49–51°C) with its toxic effects being those of HCN.[5] A concentration of 92 ppm for 10 min has caused a fatality[6] and low concentrations (10 ppm) are irritating to the eyes, nose, and respiratory tract.[5,6] It should be handled with great care only in a properly functioning chemical fume hood. Cyanogen bromide is used in laboratories for the activation of agarose beads in affinity chromatography,[7] in the analysis of protein structure,[8] and for the detection of pyridine compounds.[9]

Principles of Destruction

Inorganic cyanides and CNBr are oxidized by sodium or calcium hypochlorite,[10,11] in basic solution, to the much less toxic cyanate ion (mouse LD_{50} for sodium cyanate = 260 mg/kg[12]). Further hydrolysis of the cyanate ion is also possible.

Destruction Procedures

Destruction of Bulk Quantities[10]

A. Dissolve NaCN or CNBr in H_2O so that the concentration does not exceed 25 mg/mL for NaCN or 60 mg/mL for CNBr. Mix one volume of this solution with one volume of sodium hydroxide (NaOH) solution (1 M) and two volumes of 5.25% sodium hypochlorite [i.e., cyanide solution:NaOH:NaOCl (1:1:2)]. Stir the mixture for 3 h, test for completeness of destruction, neutralize, and discard it. Use fresh sodium hypochlorite solution (see below for assay procedure).

B. Dissolve NaCN or CNBr in H_2O so that the concentration does not exceed 25 mg/mL for NaCN or 60 mg/mL for CNBr. Mix one volume of this solution with one volume of NaOH solution (1 M) and add 60 g of calcium hypochlorite per liter of basified solution. Stir the mixture for 3 h, test for completeness of destruction, neutralize, and discard it.

Destruction of Sodium Cyanide or Cyanogen Bromide in Solution[10]

A. If necessary dilute the NaCN solution with H_2O so that the concentration does not exceed 25 mg/mL. If necessary dilute aqueous solutions of CNBr with H_2O so that the concentration does not exceed 60 mg/mL. If necessary dilute solutions of CNBr in organic solvents with the same organic solvent so that the concentrations do not exceed 60 mg/mL for acetonitrile, 30 mg/mL for dimethyl sulfoxide (DMSO), dimethylformamide (DMF), 2-methoxyethanol, or 0.1 M hydrochloric acid (HCl), 25 mg/mL for ethanol, or 19 mg/mL for N-methyl-2-pyrrolidinone. For each volume of solution add one volume of 1 M NaOH solution and two volumes of 5.25% sodium hypochlorite solution. Stir the mixture for 3 h, test for completeness of destruction, neutralize, and discard it. Use fresh sodium hypochlorite solution (see below for assay procedure).

B. If necessary dilute the NaCN solution with H_2O so that the concentration does not exceed 25 mg/mL. If necessary dilute aqueous solutions of CNBr with H_2O so that the concentration does not exceed 60 mg/mL. If necessary dilute solutions of CNBr in organic solvents with the same organic solvent so that the concentrations do not exceed 60 mg/mL for acetonitrile, 30 mg/mL for DMSO, DMF, 2-methoxyethanol, or 0.1 M HCl, 25 mg/mL for ethanol, or 19 mg/mL for N-methyl-2-pyrrolidinone. For each volume of solution add one volume of 1 M NaOH solution, then add 60 g of calcium hypochlorite per liter of basified solution. Stir the mixture for 3 h, test for completeness of destruction, neutralize, and discard it.

Destruction of Cyanogen Bromide in 70% Formic Acid[10]

A. If necessary dilute the solution so that the concentration of CNBr does not exceed 60 mg/mL and basify the solution by the **slow** addition of two volumes of 10 M potassium hydroxide (KOH) solution (a **very** exothermic process). Cool, then for each volume of solution add one volume of 1 M NaOH solution and two volumes of 5.25% sodium hypochlorite solution. Stir the mixture for 3 h, test for completeness of destruction, neutralize, and discard it. Use fresh sodium hypochlorite solution (see below for assay procedure).

B. If necessary dilute the solution so that the concentration of CNBr does not exceed 60 mg/mL and basify the solution by the **slow** addition of two volumes of 10 M KOH solution (a **very** exothermic process). Cool, then for each volume of solution add one volume of 1 M NaOH solution, then add 60 g of calcium hypochlorite per liter of basified solution. Stir the mixture for 3 h, test for completeness of destruction, neutralize, and discard it.

Destruction of Hydrogen Cyanide[11]

Dissolve the HCN in several volumes of ice water and add one molar equivalent of NaOH solution at 0–10°C. (Do **not** add NaOH, NaCN, or any base to liquid HCN; a violent reaction may occur). Add a 50% excess of 5.25% sodium hypochlorite solution (80 mL of solution for each gram of HCN) at 0–10°C with stirring and allow the mixture to warm to room temperature. After standing for several hours test for completeness of destruction, neutralize, and discard it. Use fresh sodium hypochlorite solution (see below for assay procedure).

Assay of Sodium Hypochlorite Solution

Sodium hypochlorite solutions tend to deteriorate with time so they should be periodically checked for the amount of active chlorine they contain. Pipette 10 mL of sodium hypochlorite solution into a 100-mL volumetric flask and fill to the mark with distilled H_2O. Pipette 10 mL of this solution into a conical flask containing 50 mL of distilled H_2O, 1 g of potassium iodide, and 12.5 mL of 2 M acetic acid. Titrate this solution against 0.1 N sodium thiosulfate solution using starch as an indicator. Each 1 mL of the sodium thiosulfate solution corresponds to 3.545 mg of active chlorine. The sodium hypochlorite solution used in these degradation reactions should contain 25–30 g of active chlorine/L.

Analytical Procedures[10]

Prepare the following solutions. The buffer solution was prepared by dissolving 13.6 g of potassium phosphate, monobasic (KH_2PO_4), 0.28 g of sodium phosphate, dibasic (Na_2HPO_4), and 3.0 g of potassium bromide (KBr) in distilled H_2O and making up to 1 L with distilled H_2O. The reagent was prepared by stirring 3.0 g of barbituric acid in 10 mL of H_2O and adding 15 mL of 4-methylpyridine and 3 mL of concentrated HCl while continuing to stir. After cooling the solution was diluted to 50 mL with H_2O. The sodium ascorbate solution was 10 mg/mL in H_2O, the NaCN solution was 100 mg/L in H_2O, and the Chloramine-T solution was 100 mg/mL in H_2O. The sodium ascorbate and Chloramine-T solutions were prepared fresh daily and the standard NaCN solution was prepared fresh every week. The other solutions appeared to be quite stable.

Two 1 mL portions of the reaction mixtures were centrifuged, if necessary to remove suspended solids, and each was added to 4 mL of buffer. If an orange or yellow color appeared, sodium ascorbate solution was added dropwise until the mixtures were colorless (but no more than 2 mL should be added). One solution was spiked with 200 μL of NaCN solution, and 1 mL of Chloramine-T solution was added to each solution. The solutions were shaken and allowed to stand for 1–2 min, then 1 mL of the reagent was added and the mixtures were shaken and allowed to stand for 5 min. A blue color indicates the presence of cyanide. For complete destruction the unspiked solution should be colorless and the spiked solution should be blue. The absorbance was measured at 605 nm using 10-mm cuvettes (after centrifuging again if necessary to remove suspended solids). Appropriate standards and blanks should always be run. The limit of detection was ~ 3 μg/mL.

Mutagenicity Assays

Since CNBr or cyanide ion have not been reported to be mutagenic no studies were performed.

Related Compounds

These procedures should be applicable to cyanogen chloride, cyanogen iodide, and various inorganic cyanides although we have carried out no tests. These procedures are not applicable to organic nitriles. See procedures on page 207 for methods of disposing of these compounds.

References

1. Smyth, H.F.; Carpenter, C.P.; Weil, C.S.; Pozzani, U.C.; Striegel, J.A.; Nycum, J.S. Range-finding toxicity data: List VII. *Am. Ind. Hyg. Assoc. J.* **1969**, *30*, 470–476.

2. Hayes, W.J. The 90-dose LD_{50} and a chronicity factor as measures of toxicity. *Toxicol. Appl. Pharmacol.* **1967**, *11*, 327–335.

3. Sax, N.I; Lewis, R.J., Sr. *Dangerous Properties of Industrial Materials*, 7th ed.; Van Nostrand-Reinhold: New York, 1989; pp 1904–1905.

4. Reference 3, pp 982–983.

5. Parmeggiani, L., Ed. *Encyclopedia of Occupational Safety and Health*, 3rd ed.; International Labour Office: Geneva, 1983; Vol. 1, pp 574–577.

6. Prentiss, A.M. *Chemicals in War*; McGraw-Hill: New York, 1937; p. 174.

7. Cuatrecasas, P.; Anfinsen, C.B. Affinity chromatography. In *Methods in Enzymology*; Jakoby, W.B., Ed.; Academic Press: New York, 1977; Vol. 22, pp 345–378.

8. Huang, H.V.; Bond, M.W.; Hunkapiller, M.W.; Hood,L.E. Cleavage at tryptophanyl residues with dimethyl sulfoxide-hydrochloric acid and cyanogen bromide. In *Methods in Enzymology*; Hirs, C.H.W., Timasheff, S.N., Eds.; Academic Press: New York, 1983; Vol. 91, Part 1, pp 318–324.

9. Fuentes-Duchemin, J.; Casassas, E. Photometric determination of traces of pyridine by reaction with cyanogen bromide and 4,4'-diaminostilbene-2,2'-disulphonic acid. *Anal. Chim. Acta* **1969**, *44*, 462–466.

10. Lunn, G.; Sansone, E.B. Destruction of cyanogen bromide and inorganic cyanides. *Anal. Biochem.* **1985**, *147*, 245–250.

11. National Research Council, Committee on Hazardous Substances in the Laboratory. *Prudent Practices for Disposal of Chemicals from Laboratories;* National Academy Press: Washington, DC, 1983; pp 86–87.

12. Cerami, A.; Allen, T.A.; Graziano, J.H.; deFuria, F.G.; Manning, J.M.; Gillette, P.N. Pharmacology of cyanate. I. General effects on experimental animals. *J. Pharmacol. Exp. Ther.* **1973**, *185*, 653–666.

CYCLOSERINE

> **CAUTION!** Refer to safety considerations section on page 6 before starting any of these procedures.

Cycloserine (**I**, D-4-amino-3-isoxazolidinone, D-4-amino-3-isoxazolidone, orientomycin, Cyclomycin, Farmiserine, Miroserina, Novoserin, Closina, Farmiserina, Micoserina, Oxamycin, Oxymycin, Tisomycin, Wasserina, Seromycin) is an antibacterial, tuberculostatic drug produced by *Streptomyces orchidaceus*. It is moderately toxic and it can produce a variety of effects in humans.[1] It is a solid (mp 155–156°C) and it is soluble in water.

(I)

Principle of Destruction

Cycloserine is oxidized by potassium permanganate in 3 M sulfuric acid ($KMnO_4$ in H_2SO_4).[2] The products have not been determined. Destruction is complete and <0.4% of the original compound remains.

Destruction Procedure

Take up 100 mg of cycloserine in 20 mL of 3 M H_2SO_4 and add 0.96 g of $KMnO_4$ in portions with stirring. Stir the reaction mixture at room temperature for 18 h, then decolorize it with sodium ascorbate, and neutralize with solid sodium bicarbonate. Test for completeness of destruction and discard it.

Analytical Procedure

A modification of a previously reported spectrofluorimetric method[3] was used. A pH 8.1 buffer was prepared by dissolving 11 g of sodium tetraborate decahydrate in 950 mL of H_2O and adding 42 mL of 1 M hydrochloric acid. A 0.06% solution of 1,4-benzoquinone in 95% ethanol was also prepared. Cycloserine was determined by adding 4 mL of buffer to 100 μL of the neutralized reaction mixture followed by 100 μL of the benzoquinone solution. This mixture was heated at 100 °C for 30 min then allowed to cool for 10 min. Fluorescence was determined by using a spectrophotofluorimeter set at an excitation wavelength of 381 nm and an emission wavelength of 502 nm. A relatively high background was observed with blanks made with H_2O instead of the reaction mixture and blanks from reactions in which the cycloserine was omitted. However, the method is quite sensitive and spiking experiments showed that if any cycloserine was present it was there at a concentration of <20 μg/mL.

Mutagenicity Assays

The mutagenicity assays were carried out as described on page 4 using tester strains TA98, TA100, TA1530, and TA1535. The final reaction mixture (tested at a level corresponding to 0.5 mg of undegraded material per plate) was not mutagenic. The pure compound was toxic to the cells at levels of 1000 and 500 μg/plate but not mutagenic or toxic at lower levels (250–10 μg/plate).

Related Compounds

Potassium permanganate in 3 M H_2SO_4 is a general oxidative procedure and it should be applicable to related compounds. Full validation should, however, be carried out in each case.

References

1. Sax, N.I.; Lewis, Sr., R.J. *Dangerous Properties of Industrial Materials*, 7th ed.; Van Nostrand-Reinhold: New York, 1989; p. 1015.

2. Lunn, G; Sansone, E. B. Validated methods for degrading hazardous chemicals: Some alkylating agents and other compounds. *J. Chem. Educ.* (in press).

3. El-Sayed, L.; Mohamed, Z.H.; Wahbi, A.-A.M. Spectrophotometric and spectrofluorimetric determination of cycloserine with p-benzoquinone. *Analyst* **1986**, *111*, 915–917.

DICHLOROMETHOTREXATE, VINCRISTINE, AND VINBLASTINE

CAUTION! Refer to safety considerations section on page 6 before starting any of these procedures.

The degradation of a number of antineoplastic drugs, including dichloromethotrexate(**I**), vincristine(**II**), and vinblastine(**III**), was investigated by the International Agency for Research on Cancer (IARC).[1]

(**I**)

(II)   (III)

Dichloromethotrexate (mp 185–204°C),[2] vincristine sulfate (mp 273–278°C),[3] and vinblastine sulfate (mp 284–285°C)[4] are solids. Dichloromethotrexate is not soluble in H_2O or organic solvents, but it is soluble in dilute acid or base. Vincristine sulfate and vinblastine sulfate are soluble in H_2O, chloroform, and methanol but not in ether. Vincristine is a teratogen and produces various effects on the body,[5] and vinblastine is a teratogen and has effects on the eye, heart, and bone marrow.[6] These compounds are all used as antineoplastic drugs.

Principle of Destruction

Dichloromethotrexate, vincristine sulfate, and vinblastine sulfate are destroyed by oxidation with potassium permanganate in sulfuric acid ($KMnO_4$ in H_2SO_4).[1] Destruction is >99%. The products of these reactions have not been determined.

Destruction Procedures

Destruction of Bulk Quantities of Dichloromethotrexate, Vincristine Sulfate, and Vinblastine Sulfate

Dissolve in 3 M H_2SO_4 so that the concentration does not exceed 1 mg/mL, then add 0.5 g of $KMnO_4$ for each 10 mL of solution and stir for 2 h.

Decolorize with ascorbic acid, neutralize, check for completeness of destruction, and discard it.

Destruction of Aqueous Solutions of Dichloromethotrexate, Vincristine Sulfate, and Vinblastine Sulfate

Dilute with H_2O, if necessary, so that the concentration does not exceed 1 mg/mL, then add enough concentrated H_2SO_4 to obtain a 3 M solution and allow it to cool to room temperature. For each 10 mL of solution add 0.5 g of $KMnO_4$ and stir for 2 h. Decolorize with ascorbic acid, neutralize, check for completeness of destruction, and discard it.

Destruction of Pharmaceutical Preparations of Vincristine Sulfate and Vinblastine Sulfate Containing 1 mg of Compound, 1.275 mg of Methyl p-Hydroxybenzoate, 1.225 mg of Propyl p-Hydroxybenzoate, and 100 mg of Mannitol

Dissolve in 3 M H_2SO_4 to obtain a drug concentration of 0.1 mg/mL then, for each 10 mL of solution, add 0.5 g of $KMnO_4$, in small portions to avoid frothing, and stir for 2 h. Decolorize with ascorbic acid, neutralize, check for completeness of destruction, and discard it.

Destruction of Injectable Pharmaceutical Preparations of Dichloromethotrexate Containing 2–5% Glucose and 0.45% Saline

Dilute with H_2O so that the concentration does not exceed 2.5 mg/mL, then add enough concentrated H_2SO_4 to obtain a 3 M solution and allow it to cool to room temperature. For each 10 mL of solution add 1 g of $KMnO_4$, in small portions to avoid frothing, and stir for 1 h. Decolorize with ascorbic acid, neutralize, check for completeness of destruction, and discard it.

Destruction of Solutions of Dichloromethotrexate, Vincristine Sulfate, and Vinblastine Sulfate in Volatile Organic Solvents

Remove the solvent under reduced pressure using a rotary evaporator and take up the residue in 3 M H_2SO_4 so that the concentration does not exceed 1 mg/mL. For each 10 mL of solution add 0.5 g of $KMnO_4$ and stir for 2 h. Decolorize with ascorbic acid, neutralize, check for completeness of destruction, and discard it.

Destruction of Dimethyl Sulfoxide (DMSO) or Dimethylformamide (DMF) Solutions of Dichloromethotrexate, Vincristine Sulfate, and Vinblastine Sulfate

Dilute with H_2O so that the concentration of DMSO or DMF does not exceed 20% and the concentration of the drug does not exceed 1 mg/mL, then add enough concentrated H_2SO_4 to obtain a 3 M solution and allow it to cool to room temperature. For each 10 mL of solution add 1 g of $KMnO_4$, in portions to avoid frothing, and stir for 2 h. Decolorize with ascorbic acid, neutralize, check for completeness of destruction, and discard it.

Decontamination of Glassware Contaminated with Dichloromethotrexate, Vincristine Sulfate, and Vinblastine Sulfate

Immerse the glassware in a 0.3 M solution of $KMnO_4$ in 3 M H_2SO_4 for 2 h, then clean it by immersion in ascorbic acid solution.

Decontamination of Spills of Vincristine Sulfate and Vinblastine Sulfate

Allow any organic solvent to evaporate and remove as much of the spill as possible by HEPA vacuuming (not sweeping), then rinse the area with H_2O. Take up the rinse with absorbents and decontaminate them by reaction with a 0.3 M solution of $KMnO_4$ in 3 M H_2SO_4 for 2 h. If the color fades, add more solution. Check for completeness of decontamination by using a wipe sample moistened with H_2O. Analyze the wipe for the presence of the drug.

Decontamination of Spills of Dichloromethotrexate

Allow any organic solvent to evaporate and remove as much of the spill as possible by HEPA vacuuming (not sweeping), then rinse the area with 3 M H_2SO_4. Take up the rinse with absorbents and allow the rinse and absorbents to react with 0.3 M $KMnO_4$ solution in 3 M H_2SO_4 for 1 h. If the color fades, add more solution. Check for completeness of decontamination by using a wipe moistened with 0.1 M H_2SO_4. Analyze the wipe for the presence of the drug.

Analytical Procedures

These drugs can be analyzed by HPLC using a 25-cm reverse phase column and UV detection at 254 nm. The mobile phases which have been recommended are as follows:

Dichloromethotrexate. 5 mM Tetrabutylammonium phosphate (adjusted to pH 3.5 with phosphoric acid) : methanol : acetonitrile (66:11:22) at 1.5 mL/ min **or** 20 mM ammonium formate : methanol (65:35) at 1 mL/min.

Vincristine sulfate. 5 mM Tetrabutylammonium phosphate (adjusted to pH 3.5 with phosphoric acid) : acetonitrile : tetrahydrofuran (THF) (54:26:20) at 1.5 mL/min **or** 20 mM ammonium formate : methanol (12:88) at 1 mL/min.

Vinblastine sulfate. 5 mM Tetrabutylammonium phosphate (adjusted to pH 3.5 with phosphoric acid) : acetonitrile : THF (54:26:20) at 1.5 mL/min **or** 20 mM ammonium formate : methanol (10:90) at 1 mL/min.

Mutagenicity Assays

In the IARC study[1] tester strains TA98, TA100, TA1530, and TA1535 of *Salmonella typhimurium* were used with and without metabolic activation (not all strains were used for each drug). Generally, the reaction mixtures were not mutagenic although degradation of pharmaceutical preparations of dichloromethotrexate gave two to three times the background activity with TA1530.

Related Compounds

Potassium permanganate in H$_2$SO$_4$ is a general oxidative method and should, in principle, be applicable to many drugs. However, any new application should be thoroughly validated both for complete destruction of the compound and for the production of nonmutagenic reaction mixtures.

References

1. Castegnaro, M.; Adams, J.; Armour, M-. A.; Barek, J.; Benvenuto, J.; Confalonieri, C.; Goff, U.; Ludeman, S.; Reed, D.; Sansone, E. B.; Telling, G., Eds. *Laboratory Decon-*

tamination and Destruction of Carcinogens in Laboratory Wastes: Some Antineoplastic Agents; International Agency for Research on Cancer: Lyon, 1985 (IARC Scientific Publications No. 73).

2. Other names are 3′,5′-dichloroamethopterin, 4-amino-10-methyl-3′,5′-dichloropteroylglutamic acid, 3′5′-dichloro-4-amino-4-deoxy-N^{10}-methylpteroglutamic acid, and N-[3,5-dichloro-4-[[(2,4-diamino-6-pteridinyl)methyl]methylamino]benzoyl]glutamic acid.

3. Other names are 22-oxovincaleukoblastine, leurocristine, VCR, and LCR.

4. Other names are vincaleukoblastine, VLB, and nincaluicolflastine.

5. Sax, N.I; Lewis, R.J., Sr. *Dangerous Properties of Industrial Materials*, 7th ed.; Van Nostrand-Reinhold: New York, 1989; pp 2115–2116.

6. Reference 5, p. 3469.

DIMETHYL SULFATE AND RELATED COMPOUNDS

> **CAUTION!** Refer to safety considerations section on page 6 before starting any of these procedures.

Dimethyl sulfate [DMS, sulfuric acid dimethyl ester, methyl sulfate, $(CH_3)_2SO_4$] is a clear, oily, high-boiling (bp 188°C) liquid. It is quite volatile and has no characteristic odor.[1] It is highly toxic,[2] causes severe burns and injury to the lungs, kidneys, and liver. It causes cancer in laboratory animals and may be a human carcinogen.[3-8] It is slightly soluble in H_2O (2.8%). It is used as an alkylating agent industrially and in the laboratory.

Diethyl sulfate [DES, sulfuric acid diethyl ester, ethyl sulfate, $(C_2H_5)_2SO_4$] is also a volatile liquid (bp 209°C) with a peppermint odor. It is almost insoluble in H_2O. It is used as an alkylating agent industrially and in the laboratory. Diethyl sulfate causes cancer in experimental animals and may be a human carcinogen.[9-10] It is a severe skin and eye irritant.[11]

Methyl methanesulfonate (MMS, methyl mesylate, methanesulfonic acid methyl ester, $CH_3SO_2OCH_3$) is a volatile liquid (bp 203°C), soluble to the extent of ~ 1:5 in H_2O. Methyl methanesulfonate causes cancer in experimental animals.[12]

Ethyl methanesulfonate (EMS, methanesulfonic acid ethyl ester, ethyl mesylate, $CH_3SO_2OC_2H_5$) is a volatile liquid (bp 213°C), which is at least somewhat soluble in H_2O. Ethyl methanesulfonate causes cancer in experimental animals.[13]

Butadiene diepoxide (**I**, BDE, erythritol anhydride, diepoxybutane, butadiene dioxide, Bioxiran, 2,2'-bioxirane, dianhydrothreitol, dianhydroerythritol, 1,1'-biethylene oxide, bp 56–58°C at 25 mm Hg) is miscible with H_2O. It is used industrially, particularly in the polymer industry. It causes cancer in experimental animals.[14]

1,3-Propane sultone (**II**, PS, 1,2-oxathiolane 2,2-dioxide, 3-hydroxy-1-propanesulfonic acid sultone, mp 31–33°C, bp 180°C at 30 mm Hg) is somewhat soluble in H_2O. It is used industrially, particularly in the detergent industry. It causes cancer in experimental animals.[15]

(I)   **(II)**

Although the toxicological properties of these compounds are not well known, by analogy with DMS they should be regarded as capable of causing lung injury and burns as well as being carcinogens. All of these compounds are mutagenic.

Principles of Destruction

Dimethyl sulfate is hydrolyzed by dilute base [sodium hydroxide (NaOH) solution (1 or 5 M), sodium carbonate (Na_2CO_3) solution (1 M), or ammonium hydroxide (NH_4OH) solution (1.5 M)] to methanol and methyl hydrogen sulfate.[16] Subsequent hydrolysis of methyl hydrogen sulfate to methanol and sulfuric acid is slow. We found, as others have,[17] that methyl hydrogen sulfate was nonmutagenic. Methyl hydrogen sulfate is a very poor alkylating agent.[18] When hydrolyzed using NH_4OH, the products are methylamine, dimethylamine, and trimethylamine. Hydrolysis destroyed DMS, a mutagenic compound, without producing other mutagenic species. The toxicity of methyl hydrogen sulfate is not well established, so appropriate steps should be taken to protect workers handling this material. In a similar fashion, DES, which is also mutagenic, can be hydrolyzed by the above reagents although the process is slower.[19] The products are presumably the analogous ethyl compounds. Ethanol is produced when the hydrolyzing agent is NaOH. Refluxing with alcoholic potassium hydroxide solution has also been reported to degrade dialkyl sulfates but validation details were not provided.[20] Methyl and ethyl methanesulfonates can be hydrolyzed with either 1 or 5 M NaOH solution. The products are methanol or ethanol and, presumably, methanesulfonic acid. Methyl methanesulfonate and EMS are mutagenic, but methanesulfonic acid is not.[19] Butadiene diepoxide and 1,3-propane sultone are hydrolyzed with either 1 or 5 M NaOH solution.[19] Theoretically, 1,3-propane sultone could reform on acidification, although we could find no evidence for this. It is probably prudent, however, not to acidify the reaction mixtures when the reaction is complete. Dimethyl sulfate, DES, MMS, and EMS can also be degraded by using a 1 M solution of sodium thiosulfate ($Na_2S_2O_3$).[21] The degradation efficiency is >99.5%. The products of the reaction have not been determined.

Destruction Procedures

Destruction of Bulk Quantities of Dimethyl Sulfate and Diethyl Sulfate[16,19]

Note. The reaction times given below gave good results in our tests. However, the reaction time may be affected by such factors as the size and shape of the flask and the rate of stirring. If two phases are apparent, this is an indication that the reaction is not complete. Stirring should be continued until the reaction mixture is homogeneous.

To accomplish destruction 10 mL of DMS or DES was added at once to a flask containing 500 mL of rapidly stirred 1 M NaOH solution, 1 M Na_2CO_3 solution, or 1.5 M NH_4OH solution. Fifteen minutes after the last of the DMS went into solution no DMS could be detected and 3 h after the last of the DES went into solution no DES could be detected. There was no apparent evolution of gas; the maximum temperature rise observed was 5°C. At the end of the reaction the mixture was neutralized, checked for completeness of destruction, and discarded.

This procedure may also be adapted for the destruction of larger quantities. Thus 100 mL of DMS was added to 1 L of 5 M NaOH solution[22] and the reaction mixture was stirred. Fifteen minutes after the last of the DMS went into solution no DMS could be detected. Similar results were obtained for DES but dissolution was much slower. No DES could be detected in solution 24 h after the addition of the DES to the base. The maximum temperature rise seen was 11°C. At the end of the reaction the mixture was neutralized, checked for completeness of destruction, and discarded.

Destruction of Bulk Quantities of Methyl Methanesulfonate, Ethyl Methanesulfonate, Butadiene Diepoxide, and 1,3-Propane Sultone[19]

A. To accomplish destruction 1 mL of the compound was added to 50 mL of 1 M NaOH solution and the reaction mixture was stirred for 1 h (PS), 6 h (MMS), 20 h (BDE), or 48 h (EMS). The reaction mixture was neutralized, checked for completeness of destruction, and discarded. These times may vary depending on such factors as the flask shape and the stirring rate. If the reaction mixture is not homogeneous, stirring should be continued until it is. The maximum temperature rise observed was 3°C (for PS).

B. To accomplish destruction 1 mL of the compound was added to 10 mL of 5 M NaOH solution and the reaction mixture was stirred for 1 h (PS), 2 h (MMS), 22 h (BDE), or 24 h (EMS). The reaction mixture was neutralized, checked for completeness of destruction, and discarded. These times may vary depending on such factors as the shape of the flask and the stirring rate. If the reaction mixture is not homogeneous, stirring should be continued until it is. The maximum temperature rise observed was 17°C (for MMS).

Destruction of Dimethyl Sulfate in Organic Solvents[16]

To accomplish destruction 1 mL of a solution of DMS in methanol, ethanol, dimethylsulfoxide (DMSO), acetone, or dimethylformamide (DMF) (1 mL of DMS in 10 mL of solvent) was shaken with four mL of 1 M NaOH solution, 1 M Na_2CO_3 solution, or 1.5 M NH_4OH solution until the mixture was homogeneous. After 15 min no DMS could be detected when the solvent was methanol, ethanol, DMSO, or DMF (<0.045%). After 1 h, no DMS could be detected when the solvent was acetone (<0.045%). The reaction mixture was neutralized, checked for completeness of destruction, and discarded.

For solvents not miscible with H_2O (toluene, p-xylene, benzene, 1-pentanol, ethyl acetate, chloroform, and carbon tetrachloride) 1 mL of a solution of DMS (1 mL of DMS in 10 mL of solvent) was added to 4 mL of 1 M NaOH, 1 M Na_2CO_3, or 1.5 M NH_4OH and the heterogeneous mixture was rapidly stirred. After 24 h no DMS could be detected in the organic layer (<0.045%). Acetonitrile solutions can also be decontaminated using the methods described. Although the solution should be homogeneous, it was found that two layers were present in some instances. After 3 h the solutions were all homogeneous and no DMS could be detected (<0.045%). We recommend stirring for 24 h to ensure complete destruction. When the reaction was complete the reaction mixture was neutralized and the layers were separated, checked for completeness of destruction, and discarded.

Destruction of Dimethyl Sulfate, Diethyl Sulfate, Methyl Methanesulfonate, and Ethyl Methanesulfonate[21]

Bulk quantities of DMS, DES, MMS, or EMS were dissolved in acetone and for each 1 mL of this solution 49 mL of 1 M sodium thiosulfate solution was added. The mixture was stirred for 2 h and discarded.

Spills of Dimethyl Sulfate[23]

The spill was covered with a 1:1:1 mixture of sodium carbonate or calcium carbonate, bentonite, and sand. The mixture was scooped up and added to a 10% NaOH solution. For each milliliter of DMS 10 mL of NaOH solution was used. This mixture was stirred for 24 h, checked for completeness of destruction, and discarded.

Analytical Procedures[7,24]

Note. If reaction mixtures that contain 5 M NaOH solution are to be analyzed, an aliquot of the reaction mixture should be diluted with four volumes of H_2O and 100 μL of this solution analyzed.

One hundred microliters of the solution to be analyzed was added to 1 mL of a solution of 2 mL of acetic acid in 98 mL of 2-methoxyethanol. This mixture was swirled and 1 mL of a solution of 5 g of 4-(4-nitrobenzyl)pyridine (4-NBP) in 100 mL of 2-methoxyethanol was added. The solution was heated at 100°C for 10 min then cooled in ice for 5 min. Piperidine (0.5 mL) and 2-methoxyethanol (2 mL) were added, and the violet color was determined at 560 nm using 10-mm disposable plastic cuvettes in a Gilford 240 UV/Vis spectrophotometer.

To check the efficacy of the analytical procedure, a small quantity of DMS can be added to the solution to be analyzed after the acetic acid-2-methoxyethanol has been added but before the 4-NBP is added. A positive response will indicate that the analytical technique is satisfactory. Dichloromethane interfered with the determination, giving false positives. Although we have no reason to believe that DMS was not destroyed in dichloromethane solutions (and no mutagenic responses were obtained in these cases), we were not able to verify that complete destruction occurred. On the other hand, the false positives (obtained also in blank experiments) were small, with the largest value being <0.03% after 2 days. This is equivalent to a destruction efficiency of >99.97%. We recommend carefully checking each layer of each reaction for completeness of destruction.

Using the analytical procedure described above, the limits of detection were DMS 10 mg/L, DES 27 mg/L, MMS 21 mg/L, EMS 275 mg/L, BDE 90 mg/L, and PS 66 mg/L but these could easily be reduced by increasing the volume of reaction mixture tested. For example, using 150 μL of solution containing EMS the limit of detection could be lowered to about 180 mg/L. Increasing the heating time to 60 min may also increase sensitivity.[25] We observed that a just noticeable violet color corresponded to a concentration that was about twice the detection limit given above, so the method could be used to rapidly screen a number of samples. The warming and cooling cycle may be inconvenient to carry out in many laboratories. For the analysis of DMS we found that it could be omitted provided that the solution was allowed to stand for 4 h before the piperidine and 2-methoxyethanol were added. The appropriate blanks and positive controls should always be run. When diethylamine was substituted for piperidine no noticeable decrease in sensitivity was observed.

The products from these degradation reactions were determined by gas chromatography using a Hewlett Packard HP5880 gas chromatograph equipped with a 1.8 m × 2-mm i.d. column packed with 10% Carbowax 20 M + 2% KOH on 80/100 Chromosorb W AW with flame ionization detection. The oven temperature was 150°C, the injection temperature was 200°C, the detector temperature was 300°C, and the carrier gas was nitrogen flowing at 30 mL/min. The approximate retention times were 0.5 min for methanol and 0.6 min for ethanol. The GC conditions given are only a guide and the exact conditions would have to be determined experimentally.

Mutagenicity Assays

The mutagenicity assays were carried out as described on page 4. For DMS tester strain TA100 was used and for DES, MMS, EMS, BDE, and PS tester strains TA98, TA100, TA1530, and TA1535 were used. One hundred microliters of solution (corresponding to 2.6 mg of undegraded DMS, 2.3 mg of DES, 2.5 mg of MMS, 2.3 mg of EMS, 2.0 mg of BDE, and 2.5 mg of PS) was used per plate.[19]

Solutions of DMS in various solvents were degraded using 1 M NaOH solution, 1 M Na$_2$CO$_3$ solution, or 1.5 M NH$_4$OH solution and the reaction mixtures were tested for mutagenicity. No mutagenic response was observed when solutions of DMS in methanol, H$_2$O, acetone, and DMSO were tested after 2 h of reaction, when the organic layer of solutions of DMS in dichloromethane and benzene were tested after 24 h of reaction, or when the organic layer of a solution of DMS in toluene was tested after 3 days of reaction.

The reaction mixture was cytotoxic (and therefore a determination of mutagenicity could not be made) when solutions of DMS in methanol or acetone were degraded using 1 M NaOH and when solutions of DMS in 1-pentanol were degraded using 1 M NaOH, 1 M Na$_2$CO$_3$, or 1.5 M NH$_4$OH.

A solution of DMS in ethanol was allowed to stand and gave a strong mutagenic response that slowly decreased with time, presumably as the DMS degraded. A similar decrease in activity was observed when a solution of DMS in ethanol was measured colorimetrically as described above.

No mutagenic activity was observed when DES, MMS, EMS, BDE, or PS were degraded with NaOH as described using the reaction times specified above with the exception that the degradation of BDE with 1 M NaOH solution gave reaction mixtures that were just mutagenic to TA98 without activation (59 revertants observed, control value 27). Degradation of BDE with 5 M NaOH did not give reaction mixtures that were significantly mutagenic.

Tester strains TA97, TA98, TA100, and TA102 were used when DMS, DES, MMS, and EMS were degraded using sodium thiosulfate solution.[21] No mutagenic activity was found.

Related Compounds

These destruction procedures and analytical methods should be applicable to other dialkyl sulfates, alkyl methanesulfonates, and related compounds although we have not verified this. Problems might arise when large alkyl groups are present because the compound may not be miscible with H$_2$O. In addition, large alkyl groups may slow hydrolysis to such an extent that complete destruction may not be obtained. In the work described above the hydrolysis slowed as the size of the alkyl group increased. β-

Propiolactone, which is a cyclic ester, was completely degraded using these methods, although there did seem to be some tendency to reform β-propiolactone on acidification and mutagenic reaction mixtures were produced so these procedures cannot be recommended for this compound.

Summary of Degradation Conditions for DMS, DES, MMS, EMS, BDE, and PS with Sodium Hydroxide

For full details refer to the destruction procedures given above.

Compound	Molarity of NaOH (M)	Reaction time†	Residue (%)	Maximum Temperature Rise (°C)	Products (%)
DMS	1	15 min*	<0.06	6	CH_3OH 68
	5	15 min*	<0.06	11	CH_3OH 71
DES	1	3 h*	<0.11	1	C_2H_5OH 47
	5	24 h	<0.11	1	C_2H_5OH 50
MMS	1	6 h	<0.09	1	CH_3OH 89
	5	2 h	<0.09	17	CH_3OH 65
EMS	1	48 h	<0.9	1	C_2H_5OH 88
	5	24 h	<0.9	1	C_2H_5OH 68
BDE	1	20 h	<0.4	1	
	5	22 h	<0.4	13	
PS	1	1 h	<0.24	3	
	5	1 h	<0.24	6	

†Reaction time is normally measured from initial mixing. Reaction times marked with an asterisk were measured from the time the compound had completely dissolved in the NaOH solution.

References

1. E.I. du Pont de Nemours & Co. *Dimethyl Sulfate, Properties, Uses, Storage and Handling*; Du Pont: Wilmington, DE, 1981, *Dimethyl Sulfate, Product Safety Bulletin*, Du Pont: Wilmington, DE, 1980, and *Dimethyl Sulfate, Material Safety Data Sheet*; Du Pont: Wilmington, DE, 1980.

3. International Agency for Research on Cancer. *IARC Monographs on the Evaluation of the Carcinogenic Risk of Chemicals to Humans, Supplement No. 4, Chemicals, Industrial Processes and Industries Associated with Cancer in Humans. IARC Monographs, Volumes 1 to 29*; International Agency for Research on Cancer: Lyon, 1982; pp 119–120.

4. International Agency for Research on Cancer. *IARC Monographs on the Evaluation of Carcinogenic Risk of Chemicals to Man.* Volume 4, *Some Aromatic Amines, Hydrazine and Related Substances, N-Nitroso Compounds and Miscellaneous Alkylating Agents*; International Agency for Research on Cancer: Lyon, 1974; pp 271–276.

5. Althouse, R.; Huff, J.; Tomatis, L.; Wilbourn, J. An evaluation of chemicals and industrial processes associated with cancer in humans based on human and animal data: IARC monographs volumes 1 to 20. *Cancer Res.* **1980**, *40*, 1–12.

6. Druckrey, H.; Preussmann, R.; Nashed, N.; Ivankovic, S. Carcinogene alkylierende substanzen. I. Dimethylsulfat carcinogene wirkung an ratten und wahrscheinliche ursache von berufskrebs. *Z. Krebsforsch.* **1966**, *68*, 103–111.

7. Druckrey, H.; Kruse, H.; Preussmann, R.; Ivankovic, S.; Landschütz, C. Cancerogene alkylierende substanzen. III. Alkyl-halogenide, -sulfate, -sulfonate und ringgespannte heterocyclen. *Z. Krebsforsch.* **1970**, *74*, 241–270.

8. International Agency for Research on Cancer. *IARC Monographs on the Evaluation of the Carcinogenic Risk of Chemicals to Humans, Supplement No. 7, Overall Evaluations of Carcinogenicity: An Updating of* IARC Monographs *Volumes 1 to 42*; International Agency for Research on Cancer: Lyon, 1987; pp 200–201.

9. Reference 4, pp 277–281.

10. Reference 8, p. 198.

11. Reference 2, pp 1681–1682.

12. International Agency for Research on Cancer. *IARC Monographs on the Evaluation of Carcinogenic Risk of Chemicals to Man.* Volume 7, *Some Anti-thyroid and Related Substances, Nitrofurans and Industrial Chemicals*; International Agency for Research on Cancer: Lyon, 1974; pp 253–260.

13. Reference 12, pp 245–251.

14. International Agency for Research on Cancer. *IARC Monographs on the Evaluation of Carcinogenic Risk of Chemicals to Man.* Volume 11, *Cadmium, nickel, some epoxides, miscellaneous industrial chemicals and general considerations on volatile anaesthetics*; International Agency for Research on Cancer: Lyon, 1976; pp 115–123.

15. Reference 4, pp 253–258.

16. Lunn, G.; Sansone, E.B. Validation of techniques for the destruction of dimethyl sulfate. *Am. Ind. Hyg. Assoc. J.* **1985**, *46*, 111–114.

17. Tan, E.-L; Brimer, P.A.; Schenley, R.L.; Hsie, A.W. Mutagenicity and cytotoxicity of dimethyl and monomethyl sulfates in the CHO/HGPRT system. *J. Toxicol. Environ. Health* **1983**, *11*, 373–380.

18. McCormack, W.B.; Lawes, B.C. Sulfuric and Sulfurous Esters. In *Kirk-Othmer Encyclopedia of Chemical Technology*, 3rd ed.; Wiley: New York, 1983; Vol. 22, pp 233–254.

19. Lunn, G.; Sansone, E. B. Validated methods for degrading hazardous chemicals: Some alkylating agents and other compounds. *J. Chem. Educ.* (in press).

20. National Research Council, Committee on Hazardous Substances in the Laboratory. *Prudent Practices for Disposal of Chemicals from Laboratories;* National Academy Press: Washington, DC, 1983; p. 61.

21. de Méo, M.; Laget, M.; Castegnaro, M.; Duménil, G. Evaluation of methods for destruction of some alkylating agents. *Am. Ind. Hyg. Assoc. J.* (in press).

22. Personal communication from B.I.Tobias, Chemical Safety Officer, NCI-FCRF, April 16, 1985.

23. Armour, M-.A.; Browne, L.M.; McKenzie, P.A.; Renecker, D.M.; Bacovsky,R.A., Eds. *Potentially Carcinogenic Chemicals, Information and Disposal Guide;* University of Alberta: Edmonton, Alberta, 1986; p. 60.

24. Epstein, J.; Rosenthal, R.W.; Ess, R.J. Use of γ-(4-nitrobenzyl)pyridine as an analytical reagent for ethylenimines and alkylating agents. *Anal. Chem.* **1955**, *27*, 1435–1439.

25. Preussmann, R.; Schneider, H.; Epple, F. Untersuchungen zum nachweis alkylierender agentien. II. Der nachweis verschiedener klassen alkylierender agentien mit einer modifikation der farbreaktion mit 4-(4-nitrobenzyl)-pyridin (NBP). *Arzneimittel-Forsch.* **1969**, *19*, 1059–1073.

DOXORUBICIN AND DAUNORUBICIN

> **CAUTION!** Refer to safety considerations section on page 6 before starting any of these procedures.

The degradation of a number of antineoplastic drugs, including doxorubicin and daunorubicin, was investigated by the International Agency for Research on Cancer (IARC).[1] Doxorubicin (**I**) (mp 204–205°C)[2] and daunorubicin (**II**) (mp 188–189°C)[3] are solids and they are soluble in H_2O and alcohols but not in organic solvents.

(**I**)   (**II**)

103

Doxorubicin[4-6] and daunorubicin[7] are mutagenic and carcinogenic in animals. Doxorubicin is a teratogen and has effects on the heart[8] and daunorubicin is a teratogen.[9] These compounds are employed as antineoplastic drugs.

Principles of Destruction

Doxorubicin and daunorubicin are destroyed by oxidation with potassium permanganate in sulfuric acid ($KMnO_4$ in H_2SO_4).[1] Destruction is >99%. The products of these reactions are unknown. Oxidation of doxorubicin and daunorubicin with sodium hypochlorite gave mutagenic products although chemical degradation was complete.[1] However, doxorubicin in urine can be completely destroyed by sodium hypochlorite followed by sodium thiosulfate without producing mutagenic residues.[10]

Destruction Procedures

Destruction of Bulk Quantities of Doxorubicin and Daunorubicin

Dissolve in 3 M H_2SO_4 so that the concentration does not exceed 3 mg/mL, then add 1 g of $KMnO_4$ for every 10 mL of solution, and stir for 2 h. Decolorize with ascorbic acid, neutralize, check for completeness of destruction, and discard it.

Destruction of Aqueous Solutions of Doxorubicin and Daunorubicin

Dilute with H_2O, if necessary, so that the concentration does not exceed 3 mg/mL, then add enough concentrated H_2SO_4 to obtain a 3 M solution, and allow it to cool to room temperature. For each 10 mL of solution add 1 g of $KMnO_4$ and stir for 2 h. Decolorize with ascorbic acid, neutralize, check for completeness of destruction, and discard it.

Destruction of Solid Pharmaceutical Preparations of Doxorubicin and Daunorubicin

Dissolve in H_2O so that the concentration does not exceed 3 mg/mL, then add enough concentrated H_2SO_4 to obtain a 3 M solution, and allow it to cool to room temperature. For each 10 mL of solution add 2 g of $KMnO_4$, in small portions to avoid frothing, and stir for 2 h. Decolorize with ascorbic acid, neutralize, check for completeness of destruction, and discard it.

Destruction of Liquid Pharmaceutical Preparations of Doxorubicin and Daunorubicin

Dilute with H_2O, if necessary, so that the concentration does not exceed 3 mg/mL, then add enough concentrated H_2SO_4 to obtain a 3 M solution, and allow it to cool to room temperature. For each 10 mL of solution add 2 g of $KMnO_4$, in small portions to avoid frothing, and stir for 2 h. Decolorize with ascorbic acid, neutralize, check for completeness of destruction, and discard it.

Destruction of Solutions of Doxorubicin and Daunorubicin in Volatile Organic Solvents

Remove the solvent under reduced pressure on a rotary evaporator and take up the residue in 3 M H_2SO_4 so that the concentration does not exceed 3 mg/mL. For each 10 mL of solution add 1 g of $KMnO_4$ and stir for 2 h. Decolorize with ascorbic acid, neutralize, check for completeness of destruction, and discard it.

Destruction of Dimethyl Sulfoxide (DMSO) Solutions of Doxorubicin and Daunorubicin

Dilute with H_2O so that the concentration of DMSO does not exceed 20% and the concentration of the drug does not exceed 3 mg/mL, then add enough concentrated H_2SO_4 to obtain a 3 M solution and allow it to cool to room temperature. For each 10 mL of solution add 2 g of $KMnO_4$ and stir for 2 h. Decolorize with ascorbic acid, neutralize, check for completeness of destruction, and discard it.

Destruction of Doxorubicin in Urine[10]

For each 4 mL of urine add 1 mL of 5.25% sodium hypochlorite solution followed by 100 mg of sodium thiosulfate. Destruction is rapid. Destroy the excess sodium hypochlorite by adding 100 mg of sodium bisulfite. Neutralize the reaction mixture, check for completeness of destruction, and discard it. Use fresh sodium hypochlorite solution (see below for assay procedure).

Decontamination of Glassware Contaminated with Doxorubicin or Daunorubicin

Immerse the glassware in a 0.3 M solution of $KMnO_4$ in 3 M H_2SO_4 for 2 h, then clean by immersion in ascorbic acid solution.

Decontamination of Spills of Doxorubicin or Daunorubicin

Allow any organic solvent to evaporate then cover the area with a 0.3 M solution of $KMnO_4$ in 3 M H_2SO_4 for 2 h. If the color fades, add more solution. Decolorize the area with ascorbic acid solution and neutralize by the addition of solid sodium carbonate.

Assay of Sodium Hypochlorite Solution

Sodium hypochlorite solutions tend to deteriorate with time so they should be periodically checked for the amount of active chlorine they contain. Pipette 10 mL of sodium hypochlorite solution into a 100 mL-volumetric flask and fill to the mark with distilled H_2O. Pipette 10 mL of this solution into a conical flask containing 50 mL of distilled H_2O, 1 g of potassium iodide, and 12.5 mL of 2 M acetic acid. Titrate this solution against 0.1 N sodium thiosulfate solution using starch as an indicator. Each 1 mL of the sodium thiosulfate solution corresponds to 3.545 mg of active chlorine. The sodium hypochlorite solution used in these degradation reactions should contain 25–30 g of active chlorine/L.

Analytical Procedures

These drugs can be analyzed by HPLC using a 25-cm reverse phase column and UV detection at 254 nm. The mobile phase was 0.01 M potassium phosphate, monobasic (KH_2PO_4) in 0.02 M phosphoric acid (H_3PO_4) : acetonitrile (45:55) flowing at 1.5 mL/min. Fluorescence detection with excitation at 470 nm and emission at 565 nm gives more sensitivity, but UV detection at 254 nm is usually satisfactory.

Mutagenicity Assays

In the IARC study[1] tester strains TA98, TA100, and TA102 of *Salmonella typhimurium* were used with and without metabolic activation. Generally the reaction mixtures were not mutagenic although degradation of doxorubicin with $KMnO_4$ gave twice the background activity with TA102.

Related Compounds

Potassium permanganate in H_2SO_4 is a general oxidative method and should, in principle, be applicable to many drugs. However, any new application should be thoroughly validated both for complete destruction of the compound and for the production of nonmutagenic reaction mixtures.

References

1. Castegnaro, M.; Adams, J.; Armour, M-. A.; Barek, J.; Benvenuto, J.; Confalonieri, C.; Goff, U.; Ludeman, S.; Reed, D.; Sansone, E. B.; Telling, G., Eds. *Laboratory Decontamination and Destruction of Carcinogens in Laboratory Wastes: Some Antineoplastic Agents*; International Agency for Research on Cancer: Lyon, 1985 (IARC Scientific Publications No. 73).

2. Other names are 10-[(3-amino-2,3,6-trideoxy-α-L-lyxo-hexopyranosyl)oxy]-7,8,9,10-tetrahydro-6,8,11-trihydroxy-8-(hydroxyacetyl)-1-methoxy-5,12-naphthacenedione, Adriablastima, adriamycin, 14-hydroxydaunomycin, and 14-hydroxydaunorubicine.

3. Other names are 8-acetyl-10-[(3-amino-2,3,6-trideoxy-α-L-lyxo-hexopyranosyl)oxy]-7,8,9, 10-tetrahydro-6,8,11-trihydroxy-1-methoxy-5,12-naphthacenedione, daunomycin, leukaemomycin C, rubidomycin, and Cerubidin.

4. International Agency for Research on Cancer. *IARC Monographs on the Evaluation of Carcinogenic Risk of Chemicals to Man*. Volume 10, *Some naturally occurring substances*; International Agency for Research on Cancer: Lyon, 1975; pp 43–49.

5. International Agency for Research on Cancer. *IARC Monographs on the Evaluation of the Carcinogenic Risk of Chemicals to Humans, Supplement No. 7, Overall Evaluations of Carcinogenicity: An Updating of* IARC Monographs *Volumes 1 to 42*; International Agency for Research on Cancer: Lyon, 1987; pp 82–83.

6. International Agency for Research on Cancer. *IARC Monographs on the Evaluation of the Carcinogenic Risk of Chemicals to Humans, Supplement No. 4, Chemicals, Industrial Processes and Industries Associated with Cancer in Humans. IARC Monographs, Volumes 1 to 29*; International Agency for Research on Cancer: Lyon, 1982; pp 29–31.

7. Reference 4, pp 145–152.

8. Sax, N.I; Lewis, R.J., Sr. *Dangerous Properties of Industrial Materials*, 7th ed.; Van Nostrand-Reinhold: New York, 1989; pp 91–92.

9. Reference 8, pp 1025–1026.

10. Monteith, D.K.; Connor, T.H.; Benvenuto, J.A.; Fairchild, E.J.; Theiss, J.C. Stability and inactivation of mutagenic drugs and their metabolites in the urine of patients administered antineoplastic therapy. *Environ. Mol. Mutagenesis* **1987**, *10*, 341–356

DRUGS CONTAINING HYDRAZINE AND TRIAZENE GROUPS

CAUTION! Refer to safety considerations section on page 6 before starting any of these procedures.

The drugs considered in this section all have N-N bonds either as part of a triazene group (Dacarbazine) or a hydrazine group (others). These compounds are all solids and are all at least somewhat soluble in H_2O. The basic nature of these drugs makes them very soluble in acid solution. Procarbazine is supplied and used as the hydrochloride.

The compounds considered are:

Compound Name	Melting Point (°C)	Compound Number
Procarbazine hydrochloride[1]	223–226	(I·HCl)
Isoniazid[2]	171–173	(II)
Iproniazid phosphate[3]	180–182	(III·H$_3$PO$_4$)
Dacarbazine[4]	250–255	(IV)

$CH_3NHNHCH_2$—⟨◯⟩—$CONHCH(CH_3)_2$

(I)

N⟨◯⟩—$CONHNH_2$ N⟨◯⟩—$CONHNHCH(CH_3)_2$

(II)   **(III)**

$CONH_2$ ... $N=N-N(CH_3)_2$ (imidazole ring)

(IV)

Dacarbazine is mutagenic[5] and also carcinogenic in laboratory animals[6–9] and probably carcinogenic to humans.[7] Procarbazine[10–14] and isoniazid[11,15–17] are carcinogenic in laboratory animals and procarbazine is probably carcinogenic in humans.[12] Procarbazine is a teratogen,[18] isoniazid is a teratogen and produces numerous systemic effects,[19] iproniazid is a teratogen and produces various effects including changes in liver function,[20] and dacarbazine causes nausea and has effects on the blood.[21] Procarbazine and dacarbazine are antineoplastics, isoniazid is a tuberculostatic, and iproniazid is a monoamine oxidase inhibitor.

Principles of Destruction

When these drugs are degraded by reduction with nickel-aluminum (Ni-Al) alloy in potassium hydroxide (KOH) solution, <0.03% of the original amount of dacarbazine, <0.65% procarbazine, <0.2% isoniazid, and <0.2% iproniazid remains. The products are 4-aminoimidazole-5-carboxamide and dimethylamine from dacarbazine, N-isopropyl-α-amino-p-toluamide, N-isopropyl-p-toluamide, and methylamine from procarbazine, and 4-piperidinecarboxamide from isoniazid and iproniazid. Initially, reduction of the pyridine rings of isoniazid and iproniazid gives the corresponding piperidino hydrazines but reduction for the times specified gives full reduction to 4-piperidinecarboxamide. Destruction of iproniazid also produces isopropylamine. Dacarbazine is generally used in a citric acid solution and this solution will degrade and turn pink upon exposure to light. Such degraded solutions can also be successfully treated with Ni-Al alloy.[5]

Although potassium permanganate in sulfuric acid ($KMnO_4$ in H_2SO_4) oxidation gave complete destruction of dacarbazine in all cases, the reaction mixtures were mutagenic[5] and the carcinogen N-nitrosodimethylamine was produced. Increasing the ratio of oxidant to substrate gave nonmutagenic reaction mixtures, but the volumes became too large for a practical destruction procedure. Exposure to sunlight gave only partial degradation of the dacarbazine. Potassium permanganate and calcium hypochlorite (but **not** sodium hypochlorite) can be used to degrade procarbazine. The drug is completely destroyed (<0.8% remains), the reaction mixtures are nonmutagenic, and nitrosamines are not detected.[22] The products of these reactions are not known.

Destruction Procedures

Destruction of Dacarbazine and Isoniazid

Dissolve bulk quantities in H_2O so that the concentration does not exceed 10 mg/mL. Open capsules to ensure dissolution of the drug. Dilute aqueous solutions and pharmaceutical preparations with H_2O, if necessary, so that the concentration does not exceed 10 mg/mL. Add an equal volume of 1 M KOH solution, then add 1 g of Ni-Al alloy for each 20 mL of the basified solution. Perform the reaction in a container that is at least three times as large as the final reaction volume. Add quantities of Ni-Al alloy in excess of 5 g in portions over the course of at least 1 h to avoid frothing. Stir the mixture overnight, then filter it through a pad of Celite. Neutralize the filtrate, check for completeness of destruction, and discard it. Allow the spent nickel to dry on a metal tray away from flammable solvents for 24 h, then discard it with the solid waste.

Destruction of Iproniazid

Dissolve bulk quantities in H_2O so that the concentration does not exceed 5 mg/mL. Open capsules to ensure dissolution of the drug. Dilute aqueous solutions and pharmaceutical preparations with H_2O, if necessary, so that the concentration does not exceed 5 mg/mL. Add an equal volume of 1 M KOH solution and then add 1 g of Ni-Al alloy for each 20 mL of the basified solution. Perform the reaction in a container that is at least three times as large as the final reaction volume. Add quantities of Ni-Al alloy in excess of 5 g in portions over the course of at least 1 h to avoid frothing. Stir the mixture for 96 h, then filter it through a pad of Celite. Neutralize the filtrate, check for completeness of destruction, and discard it. Allow the spent nickel to dry on a metal tray away from flammable solvents for 24 h, then discard it with the solid waste.

Destruction of Procarbazine

A. Dissolve bulk quantities in H_2O so that the concentration does not exceed 10 mg/mL. Open capsules to ensure dissolution of the drug. Dilute aqueous solutions and pharmaceutical preparations with H_2O, if necessary, so that the concentration does not exceed 10 mg/mL. Add an equal volume of 1 M KOH solution, then add 1 g of Ni-Al alloy for each 20 mL of the basified solution. Perform the reaction in a container that is at least three times as large as the final reaction volume. Add quantities of Ni-Al alloy in excess of 5 g in portions over the course of at least 1 h to avoid frothing. Stir the mixture overnight, then filter it through a pad of Celite. Neutralize the filtrate, check for completeness of destruction, and discard it. Allow the spent nickel to dry on a metal tray away from flammable solvents for 24 h, then discard it with the solid waste.

B. Take up bulk quantities in H_2O so that the concentration does not exceed 2.5 mg/mL and dilute aqueous solutions, if necessary, so that the concentration does not exceed 2.5 mg/mL. For each 10 mL of solution add 1.5 g of calcium hypochlorite and stir the mixture overnight, add ascorbic acid to destroy excess oxidant, analyze for completeness of destruction, and discard it.

C. Take up bulk quantities in H_2O so that the concentration does not exceed 5 mg/mL and dilute aqueous solutions, if necessary, so that the concentration does not exceed 5 mg/mL. For each 5 mL of solution add 1 mL of concentrated H_2SO_4 and 0.29 g of $KMnO_4$ so that the solution is 3 M in H_2SO_4 and 0.3 M in $KMnO_4$. Stir the mixture overnight, decolorize with ascorbic acid, neutralize, analyze for completeness of destruction, and discard it.

Analytical Procedures

Analysis was by HPLC using a 250 × 4.6-mm i.d. column of Microsorb C8. The injection volume was 20 μL and UV detection was at 254 nm. The mobile phase was a mixture of methanol and buffer flowing at 1 mL/min. The buffer contained 0.04% ammonium phosphate, monobasic [$(NH_4)H_2PO_4$] and 0.1% triethylamine. The methanol:buffer ratios were 10:90 for isoniazid; 20:80 for dacarbazine and 4-aminoimidazole-5-carboxamide; 50:50 for procarbazine, iproniazid, and N-isopropyl-p-toluamide; and 80:20 for N-isopropyl-α-amino-p-toluamide. On our equipment these mobile phase combinations were found to give reasonable retention times (3.4–8.3 min).

Gas chromatography using a 1.8 m × 2-mm i.d. glass column packed with 10% Carbowax 20 M + 2% KOH on 80/100 Chromosorb W AW was employed to determine the products of these reactions. The injection temperature was 200°C and the flame ionization detector operated at 300°C. The oven temperature was 60°C and approximate retention times were methylamine (0.7 min), dimethylamine (0.7 min), and isopropylamine (0.8 min). At 100°C the retention time of N-nitrosodimethylamine was 6.6 min. Using a 1.8 m × 2-mm i.d. glass column packed with 2% Carbowax 20 M + 1% KOH on 80/100 Supelcoport and an oven temperature of 200°C the products of the reductions of isoniazid and iproniazid were detected. The piperidino hydrazines from isoniazid and iproniazid had retention times of 11.8 and 8.2 min, respectively, and 4-piperidinecarboxamide had a retention time of 10.5 min.

Mutagenicity Assays[5]

The mutagenicity assays were carried out as described on page 4 using tester strains TA98, TA100, TA1530, and TA1535. The final reaction mixtures (tested at a level corresponding to 0.5 mg (0.25 mg for iproniazid and the calcium hypochlorite procedures) undegraded material per plate) were not mutagenic. Of the original compounds, only dacarbazine was found to be mutagenic and none of the products detected were found to be mutagenic.

Related Compounds

The Ni-Al alloy procedure should be applicable for related compounds containing hydrazine and triazene groups, but it should be fully validated before routine use. Nickel-aluminum alloy has been shown to reduce 3-methyl-1-p-tolyltriazene to its parent amines in good yield although the complete disappearance of the starting material has not been established.[23]

References

1. Other names are N-(1-methylethyl)-4-[(2-methylhydrazino)methyl]benzamide, ibenzmethyzin, N-4-isopropylcarbamoylbenzyl-N'-methylhydrazine, 2-(p-isopropylcarbamoylbenzyl)-1-methylhydrazine, N-isopropyl-α-(2-methylhydrazino)-p-toluamide, matulane, N-isopropyl-p-[(2-methylhydrazino)methyl]benzamide, p-(N^1-methylhydrazinomethyl)-N-isopropylbenzamide, 1-methyl-2-[(isopropylcarbamoyl)benzyl]hydrazine, natulan, and MIH. This compound is generally supplied and used as the hydrochloride.

2. Other names are 4-pyridinecarboxylic acid hydrazide, isonicotinyl hydrazide, isonicotinoyl hydrazide, isonicotinylhydrazine, isonicotinoylhydrazine, isonicotinic acid hydrazide, INH, rimitsid, Cedin(Aerosol), Isocid, Neoxin, Hidrasonil, Ertuban, Antimicina, Hyzyd, Isonex, Unicozyde, Zonazide, Hycozid, Niconyl, Isonicazide, Isonicid, Isonicotan, Tubazid, Tibizide, Isobicina, Isozide, Isonilex, Isonindon, Isotebezid, Nicetal, Nikozid, Nitadon, Nyscozid, Pelazid, Raumanon, Retozide, RU-EF-Tb, Tebecid, Tisiodrazida, Tizide, Isozyd, Sauterazid, Niplen, TB-Vis, Tekazin, Isidrina, Hydrazid, Nevin, Cotinazin, Dinacrin, Ditubin, Mybasan, Neoteben, Niadrin, Nicozide, Nydrazid, Nidaton, Nicizina, Nicotibina, Pycazide, Pyricidin, Isolyn, Pyrizidin, Rimifon, Robisellin, Isonizide, Neumandin, Isocotin, Tubicon, Tyvid, Tisin, Tibinide, Tubilysin, Tubomel, Tubeco, Atcotibine, Vazadrine, Vederon, Isdonidrin, and Zinadon.

3. Other names are 4-pyridinecarboxylic acid 2-(1-methylethyl)hydrazine, 1-isonicotinoyl-2-isopropylhydrazine, 1-isonicotinyl-2-isopropylhydrazine, N-isopropyl isonicotinhydrazide, Euphozid, Marsilid, and isonicotinic acid 2-isopropylhydrazide. This compound is frequently supplied as the phosphate salt.

4. Other names are 5-(3,3-dimethyl-1-triazenyl)-1H-imidazole-4-carboxamide, 4-(3,3-dimethyl-1-triazeno)imidazole-5-carboxamide, 4-(5)-(3,3-dimethyl-1-triazeno)imidazole-5(4)-carboxamide, 5-(3,3-dimethyl-1-triazeno)imidazole-4-carboxamide, DIC, DTIC, DTIC-Dome, and Deticene.

5. Lunn, G.; Sansone, E.B. Reductive destruction of dacarbazine, procarbazine hydrochloride, isoniazid, and iproniazid. *Am. J. Hosp. Pharm.* **1987**, *44*, 2519–2524.

6. Skibba, J.L.; Ertürk, E.; Bryan, G.T. Induction of thymic lymphosarcoma and mammary adenocarcinomas in rats by oral administration of the antitumor agent, 4(5)-(3,3-dimethyl-1-triazeno)imidazole-5(4)-carboxamide. *Cancer* **1970**, *26*, 1000–1005.

7. International Agency for Research on Cancer. *IARC Monographs on the Evaluation of the Carcinogenic Risk of Chemicals to Humans.* Volume 26, *Some Antineoplastic and Immunosuppressive Agents*; International Agency for Research on Cancer: Lyon, 1981; pp 203–215.

8. International Agency for Research on Cancer. *IARC Monographs on the Evaluation of the Carcinogenic Risk of Chemicals to Humans, Supplement No. 4, Chemicals, Industrial Processes and Industries Associated with Cancer in Humans. IARC Monographs, Volumes 1 to 29*; International Agency for Research on Cancer: Lyon, 1982; pp 103–104.

9. International Agency for Research on Cancer. *IARC Monographs on the Evaluation of the Carcinogenic Risk of Chemicals to Humans, Supplement No. 7, Overall Evaluations of Carcinogenicity: An Updating of* IARC Monographs *Volumes 1 to 42*; International Agency for Research on Cancer: Lyon, 1987; pp 184–185.

10. Kelly, M.G.; O'Gara, R.W.; Yancey, S.T.; Botkin, C. Induction of tumors in rats with procarbazine hydrochloride. *J. Natl. Cancer Inst.* **1968**, *40*, 1027–1051.

11. Kelly, M.G.; O'Gara, R.W.; Yancey, S.T.; Gadekar, K.; Botkin, C.; Oliverio, V.T. Comparative carcinogenicity of *N*-isopropyl-α-(2-methylhydrazino)-*p*-toluamide.HCl (procarbazine hydrochloride), its degradation products, other hydrazines, and isonicotinic acid hydrazide. *J. Natl. Cancer Inst.* **1969**, *42*, 337–344.

12. Reference 7, pp 311–339.

13. Reference 8, pp 220–221.

14. Reference 9, pp 327–328.

15. Reference 9, pp 227–228.

16. Reference 8, pp 146–148.

17. International Agency for Research on Cancer. *IARC Monographs on the Evaluation of Carcinogenic Risk of Chemicals to Man.* Volume 4, *Some Aromatic Amines, Hydrazine and Related Substances, N-Nitroso Compounds and Miscellaneous Alkylating Agents*; International Agency for Research on Cancer: Lyon, 1974; pp 159–172.

18. Sax, N.I; Lewis, R.J., Sr. *Dangerous Properties of Industrial Materials*, 7th ed.; Van Nostrand-Reinhold: New York, 1989; pp 2876–2877.

19. Reference 18, pp 2034–2035.

20. Reference 18, p. 2036.

21. Reference 18, pp 1022–1023.

22. Castegnaro, M.; Brouet, I.; Michelon, J.; Lunn, G.; Sansone, E.B. Oxidative destruction of hydrazines produces *N*-nitrosamines and other mutagenic species. *Am. Ind. Hyg. Assoc. J.* **1986**, *47*, 360–364.

23. Lunn, G.; Sansone, E.B.; Keefer, L.K. General cleavage of N-N and N-O bonds using nickel/aluminum alloy. *Synthesis* **1985**, 1104–1108.

ETHIDIUM BROMIDE

CAUTION! Refer to safety considerations section on page 6 before starting any of these procedures.

Ethidium bromide[1] (EB) **(I)** is a red solid (mp 260–262°C), a potent mutagen,[2] and moderately toxic.[3] It is a dye widely used in biomedical laboratories for visualizing nucleic acids. Fluorescent complexes are formed by intercalation and these complexes are readily seen on irradiation with UV light.

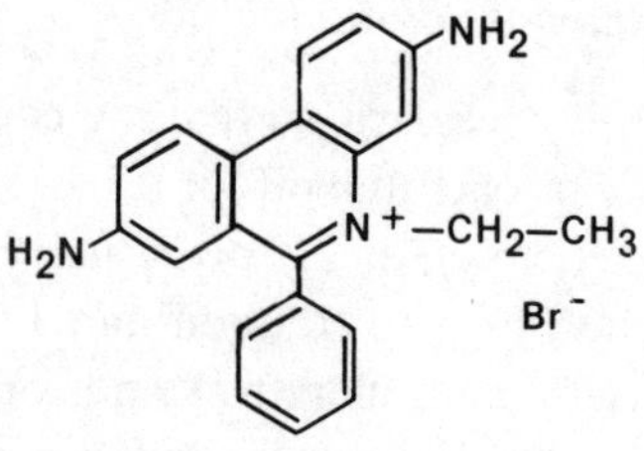

(I)

117

Principles of Destruction and Decontamination

Ethidium bromide in water and buffer solution may be degraded by reaction with sodium nitrite and hypophosphorous acid in aqueous solution.[4] This procedure may also be used to decontaminate equipment.[5] The products have not been determined but appear to consist of compounds in which the amino groups have been removed and replaced, at least partially, with oxygen.[6] Destruction efficiency is >99.87% and the resulting reaction mixtures are nonmutagenic. A modification of this method may be used to degrade EB dissolved in alcohols. The reaction is either one phase (isopropanol saturated with cesium chloride) or two phase (isoamyl alcohol or 1-butanol). Destruction efficiency is >99.75% and, in general, the reaction mixtures are not mutagenic.[7] Ethidium bromide may also be removed from solution by adsorption onto Amberlite XAD-16 resin.[4,8] Removal is >99.95% in most cases. "Blue cotton" (also called Mutasorb) has also been found to remove EB from solution[9] but it is much more expensive and much less efficient than Amberlite XAD-16 so it is not recommended.[4] Potassium permanganate and sodium hypochlorite oxidation and nickel-aluminum alloy reduction produced mutagenic reaction mixtures.[4] Potassium permanganate in hydrochloric acid has been reported to give complete destruction and nonmutagenic reaction mixtures,[10] but we have found[6] that this procedure sometimes gives mutagenic reaction mixtures.

The destruction of EB in aqueous or isoamyl alcohol solution with ozone has recently been described.[11] Destruction efficiency was >99.95% and the final reaction mixtures were nonmutagenic. Ozone is a powerful oxidizing agent and is incompatible with many organic compounds. Further investigation may be required before this procedure can be used to decontaminate isoamyl alcohol solutions on a routine basis.

Destruction and Decontamination Procedures

Destruction of EB in Aqueous Solution

A. Dilute the solution, if necessary, so that the concentration of EB does not exceed 0.5 mg/mL. For each 100 mL of EB in H_2O, MOPS buffer (see below), TBE buffer (see below), or 1 g/mL cesium chloride solution add 20 mL of 5% hypophosphorous acid solution and 12 mL of 0.5 M sodium nitrite solution, stir briefly and allow to stand for 20 h. Neutralize with sodium bicarbonate ($NaHCO_3$), check for completeness of destruction, and discard it. To prepare the hypophosphorous acid solution, add 10 mL of the commercially available 50% solution to 90 mL of H_2O and stir briefly. Prepare both the hypophosphorous acid solution and the sodium nitrite solution fresh each day.

B. **Caution! Ozone is an irritant. This reaction should be carried out in a properly functioning chemical fume hood.** Dilute the solution, if necessary, so that the concentration of EB in H_2O, Tris buffer, MOPS buffer, or cesium chloride solution does not exceed 0.4 mg/mL. Add hydrogen peroxide (H_2O_2) solution so that the concentration of H_2O_2 in the solution to be decontaminated is 1%. Pass air containing 300–400 ppm of ozone (from an ozone generator) through the solution at a rate of 2 L/min. The red solution will turn light yellow.[11] The destruction process typically takes 1 h. Check the reaction mixture for completeness of destruction and discard it. Degrade residual ozone by making the reaction mixture 1 M in sodium hydroxide.[12]

Decontamination of EB in Aqueous Solution

Dilute the solution, if necessary, so that the concentration of EB does not exceed 0.1 mg/mL. For each 100 mL of EB in H_2O, TBE buffer (see below), MOPS buffer (see below), or cesium chloride solution add 2.9 g of Amberlite XAD-16 resin, stir for 20 h, then filter the mixture. Place the beads, which now contain the EB, with the hazardous solid waste. Check the liquid for completeness of decontamination and discard it. An alternative procedure for solutions that are more concentrated than 0.1 mg/mL is to increase the relative amount of resin. This procedure should be fully validated before employing it on a routine basis.

Decontamination of Equipment Contaminated with EB

Wash the equipment once with a paper towel soaked in a decontamination solution consisting of 4.2 g of sodium nitrite and 20 mL of hypophosphorous acid (50%) in 300 mL of H_2O. Then wash five times with wet paper towels using a fresh towel each time. Soak all the towels in decontamination solution for 1 h, check for completeness of decontamination, and discard it. Make up the decontamination solution just prior to use.

If the decontamination solution (pH 1.8) is felt to be too corrosive for the surface to be decontaminated, then use six H_2O washes. Again, soak all towels in decontamination solution for 1 h before disposal. Glass, stainless steel, Formica, floor tile, and the filters of transilluminators have been successfully decontaminated using this technique.[5]

No change in the optical properties of the transilluminator filter could be detected even after a number of decontamination cycles using the decontamination solution.

Decontamination of EB in Isopropanol Saturated with Cesium Chloride

Dilute the solution, if necessary, so that the concentration of EB in the isopropanol saturated with cesium chloride does not exceed 1 mg/mL. For each volume of EB solution add four volumes of a decontamination solution consisting of 4.2 g of sodium nitrite and 20 mL of hypophosphorous acid (50%) in 300 mL of H_2O and stir the mixture for 20 h, then neutralize with $NaHCO_3$, test for completeness of destruction, and discard it.

Decontamination of EB in Isoamyl Alcohol and 1-Butanol

Dilute the solution, if necessary, so that the concentration of EB in the alcohol does not exceed 1 mg/mL. For each volume of EB solution add four volumes of a decontamination solution consisting of 4.2 g of sodium nitrite and 20 mL of hypophosphorous acid (50%) in 300 mL of H_2O and stir the two-phase mixture rapidly for 72 h. For each 100 mL of total reaction volume add 2 g of activated charcoal and stir for another 30 min. Filter the reaction mixture, neutralize with $NaHCO_3$, and separate the layers. More alcohol may tend to separate from the aqueous layer on standing. Test the layers for completeness of destruction and discard them. It should be noted that the aqueous layer contains 4.6% of 1-butanol or 2.3% of isoamyl alcohol. Discard the activated charcoal with the solid waste.

This procedure has been tested in three separate experiments for both isoamyl alcohol and 1-butanol. In one experiment one plate (TA1530 with S9 activation) indicated significant mutagenicity. The number of revertants was 2.6 times background. All the other plates for this experiment and all the other experiments showed no significant mutagenic activity. For comparison, untreated EB solutions were between 39 and 122 times background depending on the solvent and the tester strain.

Buffer Solutions

The buffer solutions used when the sodium nitrite-hypophosphorous acid method was investigated consisted of a TBE buffer (pH 8.4) containing tris(hydroxymethyl)aminomethane (0.089 M), boric acid (0.089 M), and ethylenediaminetetraacetic acid (0.002 M) and a MOPS buffer (pH 5.3) containing 4-morpholinepropanesulfonic acid (0.04 M), sodium acetate (0.0125 M), and ethylenediaminetetraacetic acid (0.00125 M).

Analytical Procedures

Although EB is itself fluorescent in solution, the complexes it forms with DNA are much more fluorescent and so the following procedure is employed. A TBE buffer is prepared containing tris(hydroxymethyl)aminomethane ($0.089 \ M$), boric acid ($0.089 \ M$), ethylenediaminetetraacetic acid ($0.002 \ M$), and sodium chloride ($0.1 \ M$). A DNA solution is prepared by dissolving 20 μg/mL of calf thymus DNA (Sigma) in the TBE buffer. A 0.1 mL aliquot of the reaction mixture is mixed with 0.9 mL of buffer, then with 1.0 mL of DNA solution. After standing for at least 15 min the fluorescence of the sample is determined using an Aminco-Bowman spectrophotofluorimeter (excitation 540 nm, emission 590 nm). Using this procedure the limit of detection of EB in the reaction mixture is ~ 0.5 μg/mL. It should be noted that this procedure only determines fluorescent compounds (such as EB) but that the EB is readily changed into nonfluorescent but still toxic compounds. Accordingly, we recommend periodic testing of reaction mixtures for mutagenicity, if possible. This testing becomes even more important if major changes are made in the procedures listed above.

EB can also be determined by thin-layer chromatography using silica gel plates eluted with 1-butanol:acetic acid:H_2O (4:1:1).[11]

Mutagenicity Assays

The mutagenicity assays were carried out as described on page 4 using tester strains TA98, TA100, TA1530, TA1535, and TA1538. When isoamyl alcohol or 1-butanol solutions were tested for mutagenicity considerable cell toxicity was seen. To avoid this, isoamyl alcohol solutions were first diluted with three volumes of dimethyl sulfoxide (DMSO) and 1-butanol solutions were first diluted with an equal volume of DMSO. Ethidium bromide itself is mutagenic only to TA98 and TA1538 with activation but, when degradation procedures other than those detailed above were employed, mutagenic activity was frequently found in other strains because the degradation procedures transformed the EB to other compounds that had different mutagenic activities. This underlines the importance of mutagenesis as well as fluorescence testing. Except as mentioned above for one experiment involving the destruction of EB in 1-butanol, all the procedures in this section produced nonmutagenic reaction mixtures.

Related Compounds

The procedures involving sodium nitrite-hypophosphorous acid and Amberlite XAD-16 may be used to deal with aqueous or buffer solutions of the related compound propidium iodide. This procedure was adapted from one recommended for the degradation of aromatic amines[13] and so it may be of use for degrading other aromatic amines (see Aromatic Amines section).

References

1. Other names are homidium bromide, Novidium bromide, Babidium bromide, 3,8-diamino-5-ethyl-6-phenylphenanthridinium bromide, 2,7-diamino-10-ethyl-9-phenylphenanthridinium bromide, and 2,7-diamino-9-phenyl-10-ethylphenanthridinium bromide.

2. MacGregor, J.T.; Johnson, I.J. In vitro metabolic activation of ethidium bromide and other phenanthridinium compounds: Mutagenic activity in *Salmonella typhimurium*. *Mutat.Res.* **1977**, *48*, 103–108.

3. Waring, M. Ethidium and propidium. In *Antibiotics*; Corcoran, J.W., Hahn, F.E., Eds.; Springer: New York, 1975; Vol. 3, pp 141–165.

4. Lunn, G.; Sansone, E.B. Ethidium bromide: Destruction and decontamination of solutions. *Anal. Biochem.* **1987**, *162*, 453–458.

5. Lunn, G.; Sansone, E.B. Decontamination of ethidium bromide spills. *Appl. Ind. Hyg.* **1989**, *4*, 234–237.

6. Lunn, G. Unpublished results.

7. Lunn, G.; Sansone, E.B. Degradation of ethidium bromide in alcohols. *BioTechniques* (in press).

8. Joshua, H. Quantitative adsorption of ethidium bromide from aqueous solutions by macroreticular resins. *BioTechniques* **1986**, *4*, 207–208.

9. Hayatsu, H.; Oka, T.; Wakata, A.; Ohara, Y.; Hayatsu, T.; Kobayashi, H.; Avimoto, S. Adsorption of mutagens to cotton bearing covalently bound trisulfo-copper-phthalocyanine. *Mutat. Res.* **1983**, *119*, 233–238.

10. Quillardet, P.; Hofnung, M. Ethidium bromide and safety - Readers suggest alternative solutions. *Trends Genet.* **1988**, *4*, 89.

11. Zocher, R.; Billich, A.; Keller, U.; Messner, P. Destruction of ethidium bromide in solution by ozonolysis. *Biol. Chem. Hoppe-Seyler* **1988**, *369*, 1191–1194.

12. Windholz, M., Ed. *The Merck Index*, 10th ed.; Merck and Co., Inc.: Rahway, NJ, 1983; p. 1002.

13. Castegnaro, M.; Barek, J.; Dennis, J.; Ellen, G.; Klibanov, M.; Lafontaine, M.; Mitchum, R.; van Roosmalen, P.; Sansone, E.B.; Sternson, L.A.; Vahl, M., Eds. *Laboratory Decontamination and Destruction of Carcinogens in Laboratory Wastes: Some Aromatic Amines and 4-Nitrobiphenyl*; International Agency for Research on Cancer: Lyon, 1985 (IARC Scientific Publications No. 64).

HALOETHERS

CAUTION! Refer to safety considerations section on page 6 before starting any of these procedures.

Chloromethylmethylether (CMME; $ClCH_2OHC_3$)[1] and bis(chloromethyl)-ether (BCME; $ClCH_2OCH_2Cl$)[2] are both colorless volatile liquids having boiling points of 59 and 104°C, respectively. Both compounds cause cancer in laboratory animals. The compound CMME may be a human carcinogen;[3-5] BCME is a human carcinogen.[4-6] Chloromethylmethylether generally contains some BCME, so conditions employed for the degradation of CMME must also degrade BCME completely. These compounds are used industrially, in organic synthesis, and may also be formed when formaldehyde and hydrogen chloride are mixed.[7]

Principles of Destruction

Both CMME and BCME can be degraded by reaction with aqueous ammonia solution (NH_3), sodium phenoxide, and sodium methoxide.[7] Degradation efficiency was >99% in all cases. Reaction with sodium phenoxide and sodium methoxide produces ethers and sodium chloride.

123

Destruction Procedures

Destruction of Bulk Quantities of CMME and BCME

A. Take up bulk quantities in acetone so that the concentration does not exceed 50 mg/mL and add an equal volume of 6% NH_3 solution. Allow the mixture to stand for 3 h, analyze for completeness of destruction, neutralize with 2 *M* sulfuric acid (H_2SO_4), and discard it.

B. Take up bulk quantities in methanol so that the concentration does not exceed 50 mg/mL. For each 1 mL of solution add 3.5 mL of a 15% (w/v) solution of sodium phenoxide in methanol. Allow this mixture to stand for 3 h, analyze for completeness of destruction, and discard it.

C. Take up bulk quantities in methanol so that the concentration does not exceed 50 mg/mL. For each 1 mL of solution add 1.5 mL of an 8–9% (w/v) solution of sodium methoxide in methanol. Allow this mixture to stand for 3 h, analyze for completeness of destruction, and discard it.

Destruction of CMME and BCME in Methanol, Ethanol, Dimethyl Sulfoxide (DMSO), Dimethylformamide (DMF), and Acetone

A. Dilute, if necessary, so that the concentration does not exceed 50 mg/mL. Add an equal volume of 6% NH_3 solution and allow the mixture to stand for 3 h, analyze for completeness of destruction, neutralize with 2 *M* H_2SO_4, and discard it.

B. Dilute, if necessary, so that the concentration does not exceed 50 mg/mL. For each 1 mL of solution add 3.5 mL of a 15% (w/v) solution of sodium phenoxide in methanol. Allow this mixture to stand for 3 h, analyze for completeness of destruction, and discard it.

C. Dilute, if necessary, so that the concentration does not exceed 50 mg/mL. For each 1 mL of solution add 1.5 mL of an 8–9% (w/v) solution of sodium methoxide in methanol. Allow this mixture to stand for 3 h, analyze for completeness of destruction, and discard it. Note that this procedure should not be used if either H_2O or chloroform (**violent reaction!**)[8] is likely to be present.

Destruction of CMME and BCME in Pentane, Hexane, Heptane, and Cyclohexane

A. Dilute, if necessary, so that the concentration does not exceed 50 mg/mL. Add an equal volume of 33% NH_3 solution and shake the mixture continuously on a mechanical shaker for at least 3 h, analyze for completeness of destruction, neutralize with 2 M H_2SO_4, and discard it.

B. Dilute, if necessary, so that the concentration does not exceed 50 mg/mL. For each 1 mL of solution add 3.5 mL of a 15% (w/v) solution of sodium phenoxide in methanol. Shake this mixture on a mechanical shaker for at least 3 h, analyze for completeness of destruction, and discard it.

C. Dilute, if necessary, so that the concentration does not exceed 50 mg/mL. For each 1 mL of solution add 1.5 mL of an 8–9% (w/v) solution of sodium methoxide in methanol. Shake this mixture on a mechanical shaker for at least 3 h, analyze for completeness of destruction, and discard it. Note that this procedure should not be used if either H_2O or chloroform (**violent reaction!**)[8] is likely to be present.

Decontamination of Glassware and Other Equipment

A. If the equipment is contaminated with CMME or BCME in a solvent that is not miscible with H_2O, rinse with acetone and treat the rinses by one of the methods described above as appropriate for treating solutions of CMME or BCME in acetone. Otherwise, immerse in a 6% NH_3 solution for at least 3 h.

B. Immerse in a 15% (w/v) solution of sodium phenoxide in methanol for at least 3 h.

C. Immerse in an 8–9% (w/v) solution of sodium methoxide in methanol for at least 3 h. Note that this procedure should not be used if either H_2O or chloroform (**violent reaction!**)[8] is likely to be present.

Decontamination of Laboratory Clothing

If the clothing is contaminated with CMME or BCME in a solvent that is not miscible with H_2O, rinse with acetone and treat the rinses by one of the methods described above as appropriate for treating solutions of CMME or BCME in acetone. Otherwise, immerse in a 6% NH_3 solution for at least 3 h.

Decontamination of Spills

Cover the spill area with an absorbent material. Saturate the absorbent material with an excess of 6% NH_3 solution. Isolate the area for at least 3 h, then neutralize with 2 M H_2SO_4 and test for completeness of destruction.

Analytical Procedures

Analysis was by gas chromatography on a 4.3 m × 2-mm i.d. glass column packed with 10% ethylene glycol adipate (EGA) on 80–100 Chromosorb W AW treated with dimethyldichlorosilane. The oven temperature was 60°C, the injection temperature was 200°C, and the detector temperature was 250°C. An electron capture detector was used. Either direct injection of the sample or injection of a 2 mL head space sample taken with a gas-tight syringe may be used. A precolumn should be used and it may be necessary to replace this after each injection.

Mutagenicity Assays

The mutagenicity assays were carried out as described on page 4 using tester strains TA100, TA1530, and TA1535. The final reaction mixtures were not mutagenic; both CMME and BCME were mutagenic.

Related Compounds

These destruction methods might be expected to be applicable to compounds of the general form ROCH₂Cl, but no tests have been carried out. Application of these methods **must** be thoroughly validated before being put into routine use.

References

1. Other names are chloromethoxymethane, methyl chloromethyl ether, monochloromethyl ether, and chlorodimethyl ether.
2. Other names are chloro(chloromethoxy)methane, *sym*-dichlorodimethyl ether, *sym*-dichloromethylether, dimethyl-1,1'-dichloroether, oxybis(chloromethane), M-chlorex, and Bis-CME.
3. International Agency for Research on Cancer. *IARC Monographs on the Evaluation of Carcinogenic Risk of Chemicals to Man.* Volume 4, *Some Aromatic Amines, Hydrazine and Related Substances,* N-*Nitroso Compounds and Miscellaneous Alkylating Agents*; International Agency for Research on Cancer: Lyon, 1974; pp 239–245.
4. International Agency for Research on Cancer. *IARC Monographs on the Evaluation of the Carcinogenic Risk of Chemicals to Humans, Supplement No. 4, Chemicals, Industrial Processes and Industries Associated with Cancer in Humans. IARC Monographs, Volumes 1 to 29*; International Agency for Research on Cancer: Lyon, 1982; pp 64–66.

5. International Agency for Research on Cancer. *IARC Monographs on the Evaluation of the Carcinogenic Risk of Chemicals to Humans, Supplement No. 7, Overall Evaluations of Carcinogenicity: An Updating of* IARC Monographs *Volumes 1 to 42*; International Agency for Research on Cancer: Lyon, 1987; pp 131–133.

6. Reference 3, pp 231–238.

7. Castegnaro, M.; Alvarez, M.; Iovu, M.; Sansone, E. B.; Telling, G.M.; Williams, D.T., Eds. *Laboratory Decontamination and Destruction of Carcinogens in Laboratory Wastes: Some Haloethers*; International Agency for Research on Cancer: Lyon, 1984 (IARC Scientific Publications No. 61).

8. Bretherick, L. *Handbook of Reactive Chemical Hazards*, 3rd ed.; Butterworths: London, 1985; p. 134.

HALOGENATED
COMPOUNDS

> **CAUTION!** Refer to safety considerations section on page 6 before starting any of these procedures.

Many hazardous compounds such as pesticides and PCBs contain halogen atoms. Little work has been done on processes suitable for the chemical degradation of these compounds in the laboratory but some work has been done on model compounds. Validated procedures are available for the following compounds.

Iodomethane	(Methyl iodide)	bp 41–43°C
2-Chloroethanol[1]		bp 129°C
2-Bromoethanol[2]		bp 56–57°C/20 mm Hg
2-Chloroethylamine hy-drochloride		mp 143–146°C
2-Bromoethylamine hydrobromide		mp 172–174°C
2-Chloroacetic acid	(Chloroethanoic acid)	mp 62–64°C

2,2,2-Trichloroacetic acid[3]		mp 54–56°C
1-Chlorobutane	(n-Butyl chloride)	bp 77–78°C
1-Bromobutane	(n-Butyl bromide)	bp 100–104°C
1-Iodobutane	(n-Butyl iodide)	bp 130–131°C
2-Bromobutane	(s-Butyl bromide)	bp 91°C
2-Iodobutane	(s-Butyl iodide)	bp 119–120°C
2-Bromo-2-methylpropane	(t-Butyl bromide)	bp 72–74°C
2-Iodo-2-methylpropane[4]		bp 99–100°C
3-Chloropyridine		bp 148°C
Chlorobenzene[5]		bp 132°C
Bromobenzene	(Phenyl bromide)	bp 156°C
Iodobenzene	(Phenyl iodide)	bp 188°C
2-Chloroaniline	(2-Chlorobenzenamine)	bp 208–210°C
3-Chloroaniline	(3-Chlorobenzenamine)	bp 230°C
2-Chloronitrobenzene		mp 33–35°C
3-Chloronitrobenzene		mp 42–44°C
4-Chloronitrobenzene		mp 83–84°C
Benzyl chloride[6]		bp 177–181°C
Benzyl bromide[7]		bp 198–199°C
α,α-Dichlorotoluene[8]		bp 82°C/10 mm Hg
1-Chlorodecane		bp 223°C
1-Bromodecane		bp 238°C

Most of these compounds are volatile liquids and the solids may also have an appreciable vapor pressure. 2-Chloroethylamine and 2-bromoethylamine generally come as their nonvolatile hydrochloride and hydrobromide salts, respectively, but the free bases would be expected to be volatile. Many of these compounds are corrosive to the skin. Iodomethane is a strong narcotic and anesthetic;[9] 2-chloroethanol is a teratogen and may affect the nervous system, liver, spleen, and lungs;[10] 2-chloroethylamine may polymerize explosively;[10] chloroacetic acid is corrosive to the skin, eyes, and mucous membranes;[11] trichloroacetic acid is corrosive and irritating to the skin, eyes, and mucous membranes;[12] 2-bromobutane is narcotic in high concentrations;[13] chlorobenzene is a teratogen[14] and its vapor may cause drowsiness and unconsciousness;[15] iodobenzene explodes on heating above 200°C;[16] 3-chloronitrobenzene is a poison and may give rise to cyanosis and blood changes;[17] benzyl chloride is a corrosive irritant to the skin, eyes, and mucous membranes and can decompose explosively under certain circumstances;[18]

and benzyl bromide[19] and α,α-dichlorotoluene[20] are lachrymators, intensely irritating to the skin, and cause CNS depression in large doses. The chloroanilines[21] and chloronitrobenzenes[22] are toxic by inhalation, skin contact, and ingestion. Halogenated compounds may react violently and explosively with alkali metals such as sodium[23] and potassium.[24] With the exception of 2-chloroethanol, 2-bromoethanol, 2-chloroethylamine, 2-bromoethylamine, 2-chloroacetic acid, 2,2,2-trichloroacetic acid, and 3-chloropyridine the compounds are not soluble in H_2O. Iodomethane is slightly soluble in H_2O. They are all soluble in alcohols and organic solvents.

Iodomethane,[25,26] benzyl chloride,[27-29] α,α-dichlorotoluene,[29,30] 2-bromoethanol,[31] 1-iodobutane,[32] 2-bromobutane,[32] 2-iodobutane,[32] 2-bromo-2-methylpropane,[32] chlorobenzene,[14] 2-chloronitrobenzene,[33] and 4-chloronitrobenzene[33] are carcinogenic in experimental animals.

These compounds are used as intermediates in industry and in organic synthesis. Iodomethane occurs naturally and is used as a chemical intermediate.

Principles of Destruction

The halogenated compounds are reductively dehalogenated with nickel-aluminum (Ni-Al) alloy in dilute base to give the corresponding compound without the halogen.[34] When the products are soluble in H_2O the yield of product is good, but when they are not (for example, toluene from benzyl chloride) they are lost from solution and accountances are not complete. The product from the chloronitrobenzenes is aniline (concomitant reduction of the nitro group) and the product from the reduction of 3-chloropyridine is piperidine in 87% yield (concomitant reduction of the initial product pyridine). In general, where the starting material can be detected by chromatography, <1% remains. 1-Bromodecane, 1-chlorodecane, and 1-chlorobutane cannot be degraded by this procedure because they are too insoluble in the aqueous methanol solvent used.

Iodomethane, benzyl chloride, benzyl bromide, 2-chloroethanol, 2- bromoethanol, 2-chloroacetic acid, 1-chlorodecane, 1-bromodecane, 1-bromobutane, 1-iodobutane, 2-bromobutane, 2-iodobutane, 2-bromo-2-methylpropane, and 2-iodo-2-methylpropane are completely degraded by refluxing them in 4.5 M ethanolic potassium hydroxide (KOH) solution for 2 h and 1-chlorobutane is completely degraded after refluxing for 4 h.[34] The products detected are the corresponding ethyl ethers. Of the compounds tested, substantial amounts of 2-chloroaniline, 3-chloropyridine, chlorobenzene, bromobenzene, and iodobenzene are left, α,α-dichlorotoluene gives incomplete destruction even after refluxing for 4 h, 2-bromoethylamine and 2-

chloroethylamine give mutagenic residues, and 2-, 3-, and 4- chloronitrobenzene are completely degraded but give products that have not yet been completely identified but appear to include azo and azoxy compounds.

Destruction Procedures[34]

With Ni-Al Alloy (not for 1-Bromodecane, 1-Chlorodecane, and 1-Chlorobutane)

Take up 0.5 mL of the halogenated compound (or 0.5 g if it is a solid) in 50 mL of H_2O (chloroacetic acid, trichloroacetic acid, 2-chloroethanol, 2-bromoethanol, 2-chloroethylamine, and 2-bromoethylamine) or methanol (other compounds) and add 50 mL of 2 M KOH solution. Stir this mixture and add 5 g of Ni-Al alloy in portions to avoid frothing. Stir the reaction mixture overnight, then filter it through a pad of Celite. Check the filtrate for completeness of destruction, neutralize, and discard it. Note that the filtrate contains the dehalogenated material. Place the spent nickel on a metal tray and allow it to dry away from flammable solvents for 24 h. Dispose of it with the solid waste.

With Ethanolic Potassium Hydroxide (not for 2-Chloroethylamine, 2-Bromoethylamine, 2,2,2-Trichloroacetic acid, 3-Chloropyridine, Chlorobenzene, Bromobenzene, Iodobenzene, α,α-Dichlorotoluene, the Chloroanilines, or the Chloronitrobenzenes)

Take up 1 mL of the halogenated compound in 25 mL of 4.5 M ethanolic KOH and reflux the mixture with stirring for 2 h (4 h for 1-chlorobutane). Cool the mixture and dilute it with at least 100 mL of H_2O. Separate the layers, if necessary, check for completeness of destruction, neutralize, and discard it. (Prepare the ethanolic KOH by dissolving 79 g of KOH in 315 mL of 95% ethanol. When volatile halides, for example, iodomethane, are to be degraded, it is important to allow this mixture to cool completely before adding the halide.)

Analytical Procedures

For analysis by gas chromatography (GC) a 1.8 m × 2-mm i.d. packed column was used together with flame ionization detection.[34] The injection temperature was 200°C, the detector temperature was 300°C, and the carrier gas was nitrogen flowing at 30 mL/min. It was frequently found advantageous to extract the final reaction mixture with 50 mL of ether or dichloromethane, and then to analyze both the aqueous layer and the extract. In this way trace amounts of compounds that were poorly soluble in H_2O could be detected. The GC conditions shown are only a guide and the exact conditions would have to be determined experimentally.

The limits of detection were in the range of 0.01–0.1 mg/mL. In each case the reaction mixture was analyzed for the presence or absence of the compound that had been degraded. If the compound was found to be absent, a small quantity of the compound was added to an aliquot of the actual reaction mixture, which was then analyzed again. The presence of an appropriate peak in the spiked reaction mixture confirmed that any halide that survived would be detected. Compounds found to be particularly sensitive to degradation on the column packings that contained base were analyzed as follows. Before analysis, 2 mL of the reaction mixture was acidified with 0.2 mL of concentrated hydrochloric acid and this mixture was neutralized by adding solid sodium bicarbonate until the effervescence of carbon dioxide ceased. This neutralized solution was analyzed using column packing A, which did not contain any base.

2-Bromoethylamine and 2,2,2-trichloroacetic acid could not be determined by GC, but their products could be determined and a good accountance for these products gave some assurance that the compounds were degraded. Products were generally determined using the same chromatographic conditions as used for the parent compounds, although it was frequently necessary to employ a lower oven temperature.

Table 1   Summary of Gas Chromatography Conditions

Compound	Packing[A]	Temperature (°)	Approximate Retention Time (min)
Iodomethane	A	60	0.5
2-Chloroethanol	A	120	1.5
2-Bromoethanol	A	120	7.8
2-Chloroethylamine	B	60	1.9
2-Chloroacetic acid	C	200	2.1
1-Chlorobutane	D	60	5.2
1-Bromobutane[B]	D	100	3.4
1-Iodobutane	E	60	3.9
2-Bromobutane	D	100	2.6
2-Iodobutane	D	100	5.4
2-Bromo-2-methylpropane	D	60	4.1

Table 1   (Continued)

Compound	Packing[A]	Temperature (°)	Approximate Retention Time (min)
2-Iodo-2-methylpropane	D	100	3.2
3-Chloropyridine	B	80	2.0
Chlorobenzene	B	60	2.0
Bromobenzene	B	60	4.1
Iodobenzene	B	60	11.2
2-Chloroaniline	B	150	1.8
3-Chloroaniline	B	150	4.2
2-Chloronitrobenzene	B	150	2.5
3-Chloronitrobenzene	B	130	3.5
4-Chloronitrobenzene	B	130	4.0
Benzyl chloride	A	120	2.7
Benzyl bromide	A	150	2.5
α,α-Dichlorotoluene	A	150	2.3
1-Chlorodecane	B	80	3.9
1-Bromodecane	B	100	1.7

[A] The column packings were

A 5% Carbowax 20 M on 80/100 Chromosorb W HP

B 2% Carbowax 20 M + 1% KOH on 80/100 Supelcoport

C 5% FFAP on 80/100 Gas Chrom Q

D 28% Pennwalt 223 + 4% KOH on 80/100 Gas Chrom R

E 10% Carbowax 20 M + 2% KOH on 80/100 Chromosorb W AW

[B] When 1-bromobutane was degraded using ethanolic KOH it was found that a small 1-butanol peak interfered with the determination of 1-bromobutane. A column packed with 20% Carbowax 20 M on 80/100 Supelcoport gave sufficient resolution. The oven temperature was 100°C, the carrier gas flowed at 20 mL/min, and the retention time was ~ 2.5 min.

Mutagenicity Assays

The mutagenicity assays were carried out as described on page 4 using tester strains TA98, TA100, TA1530, and TA1535. The final neutralized reaction mixtures were not mutagenic. This corresponds to a level of 0.5 μL (from Ni-Al alloy reactions) or 0.8 μL (from ethanolic KOH reactions) of undegraded material per plate. In some cases immiscible layers from the ethanolic KOH reactions were removed and tested directly at 1 mg/plate. None of these layers were mutagenic. Of the pure compounds tested at levels of 1–0.25 mg/plate in dimethyl sulfoxide solution, 2,2,2-trichloroacetic acid, 2-chloroethanol, 2-bromoethanol, 2-chloroethylamine, 2-bromoethylamine, 1-bromobutane, 2-bromobutane, benzyl chloride, benzyl bromide, and α,α-dichlorotoluene were mutagenic. Iodomethane has been reported to be a weak mutagen at high concentrations.[26] The products which we were able to identify were tested and none of them were found to be mutagenic. When 2-chloroethylamine and 2-bromoethylamine were treated with ethanolic KOH, 2-ethoxyethylamine was obtained, which was mutagenic to TA100, TA1530, and TA1535 with and without activation.

Related Compounds

It has been reported that many compounds can be dehalogenated with Ni-Al alloy.[35] However, although the products were identified, it was not shown that the starting materials were completely degraded. Thus, these procedures cannot be regarded as validated. Based on the results described above, however, it seems likely that Ni-Al alloy should be generally applicable to the reductive dehalogenation of halogenated compounds and should give complete destruction of the starting material. The only exceptions among the compounds studied were 1-chlorobutane, 1-chlorodecane, and 1-bromodecane, presumably because they were so insoluble in the aqueous methanol. It should be possible to degrade most halogenated compounds, which have at least some solubility in aqueous methanol. Validation must be performed before any of these compounds are routinely degraded with Ni-Al alloy. Compounds that have been reported by other workers to be reductively dehalogenated, in good yield, to the corresponding compound lacking halogen include: 4-chlorophenol,[36] 2-chlorohydroquinone,[36] 2,4,6-trichlorophenol,[37] 2,4,6-tribromophenol,[38] 5-chloroisophthalic acid,[39] 4-fluorobenzoic acid,[36] 2-chloro-5-fluorobenzoic acid,[36] 2-chlorobenzoic acid,[36] 3,4-dichlorobenzoic acid,[36] 4-bromobenzoic acid,[36] 2,4-dichlorobenzoic acid,[36] 2-chlorophenylacetic acid,[36] 4-bromophenylacetic acid,[36] 3,4-dichlorophenylacetic acid,[36] 2,4-dichlorophenylacetic acid,[36] 4-(4-chloro-3-methylphenyl)butyric acid,[36] 4(4-chloro-2-methylphenyl)butyric acid,[36] and 4-chlorophenoxyacetic acid.[36]

Refluxing in ethanolic KOH solution can also be used to degrade some halides. The procedure failed, however, for a number of compounds, particularly those with unreactive halogen atoms. Thus, it does not appear to have the wide applicability of Ni-Al alloy reduction, but it should be applicable to a number of other halides. Complete validation must be performed before the procedure is used routinely, however. When a higher ratio of organic halide:ethanolic KOH was used the reaction mixtures were frequently found to be mutagenic.[34]

Summary of Destruction Procedures

Table 2 Degradation of Halogenated Compounds Using Ni-Al Alloy

Compound	Residue (%)	Products	(%)
Iodomethane[A]	<1.1		
2-Chloroethanol	<0.36	EtOH	97
2-Bromoethanol	<0.54	EtOH	93
2-Chloroethylamine	<0.33	$EtNH_2$	100
2-Bromoethylamine		$EtNH_2$	93
2-Chloroacetic acid	<0.6	CH_3COOH	101
2,2,2-Trichloroacetic acid		CH_3COOH	104
1-Bromobutane[A]	<0.45[B]		
1-Iodobutane[A]	<0.37[B]		
2-Bromobutane[A]	<0.79[B]		
2-Iodobutane[A]	<0.66[B]		
2-Bromo-2-methylpropane[A]	<0.31[B]		
2-Iodo-2-methylpropane[A]	<0.44[B]		
3-Chloropyridine[A]	<0.30[C]	Piperidine	87
Chlorobenzene[A]	<0.14[B]	Benzene	14
Bromobenzene[A]	<0.09[B]	Benzene	17
Iodobenzene[A]	<0.14[B]	Benzene	15
2-Chloroaniline[A]	<0.06[B]	$PhNH_2$	96
3-Chloroaniline[A]	<0.14[B]	$PhNH_2$	103
2-Chloronitrobenzene[A]	<0.20[B]	$PhNH_2$	85
3-Chloronitrobenzene[A]	<0.26[B]	$PhNH_2$	82
4-Chloronitrobenzene[A]	<0.16[B]	$PhNH_2$	104
Benzyl chloride[A]	<0.07[B]	Toluene	25
Benzyl bromide[A]	<0.10[B]	Toluene	46
α,α-Dichlorotoluene[A]	<0.24[B]	Toluene	30

All reactions were carried out as described above.

[A] Reagent was initially dissolved in methanol. H_2O was the initial solvent for the other reactions.

[B] The final reaction mixture was extracted with ether or dichloromethane and each layer was separately analyzed. The results shown here are those obtained for the organic layer. No halogenated compounds were found in any of the aqueous layers. Control experiments showed that at least 74% of any of the halogenated compounds listed would be extracted into the organic layer.

[C] Pyridine was found as an intermediate but was present at <0.3% in the final reaction mixture.

Table 3 Degradation of Halogenated Compounds Using 4.5 M Potassium Hydroxide in 95% Ethanol

Compound	Residue (%)	Products	(%)
Iodomethane	<0.09		
2-Chloroethanol	<0.34	$EtOCH_2CH_2OH$	95
2-Bromoethanol	<1.25	$EtOCH_2CH_2OH$	67
2-Chloroacetic acid	<0.63	$EtOCH_2COOH$	82
1-Chlorobutane	<0.43[A,B]	BuOEt 48; BuOH	8
1-Bromobutane	<0.20[A]	BuOEt 57; BuOH	5
1-Iodobutane	<0.18[A]	BuOEt 35; BuOH	5
2-Bromobutane	<0.18[A]		
2-Iodobutane	<0.50[A]		
2-Bromo-2-methylpropane	<0.45[A]		
2-Iodo-2-methylpropane	<0.50[A]		
Benzyl chloride	<0.24[A]	$PhCH_2OEt$	85
Benzyl bromide	<0.23[A]	$PhCH_2OEt$	76
1-Chlorodecane	<0.96[A]	DecOEt	97
1-Bromodecane	<0.94[A]	DecOEt	81

All reactions were performed as described above.

[A] The final reaction mixture was extracted with ether or dichloromethane and each layer was separately analyzed. The results shown here are those obtained for the organic layer. No halogenated compounds were found in any of the aqueous layers. Control experiments showed that at least 98% of any of the halogenated compounds listed would be extracted into the organic layer.

[B] The reaction mixture was refluxed for 4 h.

References

1. Other names are 2-chloroethyl alcohol, ethylene chlorohydrin, and glycol chlorohydrin.

2. Other names are β-bromoethyl alcohol, ethylene bromohydrin, and glycol bromohydrin.

3. Other names are TCA and trichloroethanoic acid.

4. Other names are t-butyl iodide and trimethyliodomethane.

5. Other names are phenyl chloride and benzene chloride.

6. Other names are tolyl chloride, (chloromethyl)benzene, and α-chlorotoluene.

7. Other names are (bromomethyl)benzene, ω-bromotoluene, and α-bromotoluene.

8. Other names are (dichloromethyl)benzene, benzal chloride, benzyl dichloride, benzylene chloride, and benzylidene chloride.

9. Sax, N.I; Lewis, R.J., Sr. *Dangerous Properties of Industrial Materials*, 7th ed.; Van Nostrand-Reinhold: New York, 1989; p. 1999.

10. Reference 9, p. 814.

11. Reference 9, p. 773.

12. Reference 9, p. 3323.

13. Reference 9, p. 560.

14. Reference 9, pp 366–367.

15. Bretherick, L., Ed. *Hazards in the Chemical laboratory*, 4th ed.; Royal Society of Chemistry: London, 1986; pp 239–240.

16. Reference 9, p. 1996.

17. Reference 9, p. 854.

18. Reference 9, p. 3299.

19. Reference 9, p. 411.

20. Reference 9, p. 1178.

21. Reference 15, pp 238–239.

22. Reference 15, p. 248.

23. Bretherick, L. *Handbook of Reactive Chemical Hazards*, 3rd ed.; Butterworths: London, 1985; pp 1317–1318.

24. Reference 23, p. 1236.

25. International Agency for Research on Cancer. *IARC Monographs on the Evaluation of the Carcinogenic Risk of Chemicals to Man*. Volume 15, *Some Fumigants,the Herbicides 2,4-D and 2,4,5-T, Chlorinated Dibenzodioxins and Miscellaneous Industrial Chemicals*; International Agency for Research on Cancer: Lyon, 1977; pp 245–254.

26. International Agency for Research on Cancer. *IARC Monographs on the Evaluation of the Carcinogenic Risk of Chemicals to Humans*. Volume 41, *Some Halogenated Hydrocarbons and Pesticide Exposures*; International Agency for Research on Cancer: Lyon, 1986; pp 213–227.

27. International Agency for Research on Cancer. *IARC Monographs on the Evaluation of Carcinogenic Risk of Chemicals to Man*. Volume 11, *Cadmium, nickel, some epoxides, miscellaneous industrial chemicals and general considerations on volatile anaesthetics*; International Agency for Research on Cancer: Lyon, 1976; pp 217–223.

28. International Agency for Research on Cancer. *IARC Monographs on the Evaluation of the Carcinogenic Risk of Chemicals to Humans*. Volume 29, *Some Industrial Chemicals and Dyestuffs*; International Agency for Research on Cancer: Lyon, 1982; pp 49–63.

29. International Agency for Research on Cancer. *IARC Monographs on the Evaluation of the Carcinogenic Risk of Chemicals to Humans, Supplement No. 7, Overall Evaluations of Carcinogenicity: An Updating of* IARC Monographs *Volumes 1 to 42*; International Agency for Research on Cancer: Lyon, 1987; pp 148–149.

30. Reference 28, pp 65–72.

31. Theiss, J.C.; Shimkin, M.B.; Poirier, L.A. Induction of pulmonary adenomas in strain A mice by substituted organohalides. *Cancer Res.* **1979**, *39*, 391–395.

32. Poirier, L.A.; Stoner, G.D.; Shimkin, M.B. Bioassay of alkyl halides and nucleotide base analogs by pulmonary tumor response in strain A mice. *Cancer Res.* **1975**, *35*, 1411-1415.

33. Weisburger, E.K.; Russfield, A.B.; Homburger, F.; Weisburger, J.H.; Boger, E.; Van Dongen, C.G.; Chu, K.C. Testing of twenty-one environmental aromatic amines or derivatives for long-term toxicity or carcinogenicity. *J. Environ. Path. Toxicol.* **1978**, *2*, 325-356.

34. Lunn, G. Unpublished observations.

35. Keefer, L.K.; Lunn, G. Nickel-aluminum alloy as a reducing agent. *Chem. Rev.* **1989**, *89*, 459-502.

36. Buu-Hoï, N.P.; Xuong, N.D.; Bac, N.V. Déshalogénation des composés organiques halogénés au moyen des alliages de Raney (nickel ou cobalt), et ses applications à la chimie préparative et structurale. *Bull. Soc. Chim. Fr.* **1963**, 2442-2445.

37. Tashiro, M.; Fukata, G. Studies on selective preparation of aromatic compounds. 12. Selective reductive dehalogenation of some halophenols with zinc powder in basic and acidic media. *J. Org. Chem.* **1977**, *42*, 835-838.

38. Tashiro, M.; Fukata, G. The reductive debromination of bromophenols. *Org. Prep. Proced. Int.* **1976**, *8*, 231-236.

39. Märkl, G. Diensynthesen mit chlorierten α-pyronen. *Chem. Ber.* **1963**, *96*, 1441-1445.

HYDRAZINES

CAUTION! Refer to safety considerations section on page 6 before starting any of these procedures.

Hydrazines constitute a broad class of compounds having two nitrogen atoms linked together by a single bond. They are of the general form R_1R_2N-NR_3R_4, where R_1, R_2, R_3, and R_4 may be aryl or alkyl groups or hydrogen. Hydrazine itself,[1-3] 1,1-dimethylhydrazine,[4] 1,2-dimethylhydrazine,[5] methylhydrazine,[6] 1,2-diphenylhydrazine,[7] and phenyl hydrazine[8] cause cancer in experimental animals. Procarbazine[9-13] and isoniazid[10,14-16] are carcinogenic in laboratory animals and procarbazine is probably carcinogenic in humans.[11] Procarbazine,[17] methylhydrazine,[18] and iproniazid[19] are teratogens; isoniazid is a teratogen and produces a number of toxic effects in humans.[20] p-Tolylhydrazine has been reported to produce neoplastic effects.[21] Hydrazine presents an explosion hazard under certain conditions, including distillation in air,[22] and can produce skin sensitization, systemic poisoning, and may damage the liver and red blood cells,[23] methylhydrazine is corrosive to skin, eyes, and mucous membranes and may self-ignite in air,[18] 1,1-dimethylhydrazine[24] and 1,2-dimethylhydrazine[25] present a fire hazard when

141

exposed to oxidizers, and phenyl hydrazine produces a variety of toxic effects including dermatitis, anemia, and injury to the spleen, liver, bone marrow, and kidneys as well as presenting a fire hazard with oxidizers.[8] Procarbazine is an antineoplastic, isoniazid is a tuberculostatic, and iproniazid is a monoamine oxidase inhibitor. The lower molecular weight hydrazines are generally colorless liquids which are generally soluble in H_2O and organic solvents. Because of their basic nature, most hydrazines are soluble in acid. Strictly speaking those hydrazines with a carbonyl adjacent to one of the nitrogens (for example, diphenylcarbazide, isoniazid, and iproniazid) are called hydrazides but we will use the generic term hydrazine to embrace both groups here. Methods for the degradation of hydrazine, methylhydrazine, 1,1-dimethylhydrazine, 1,2-dimethylhydrazine dihydrochloride, and procarbazine have been validated by a collaborative study.[26] Some hydrazines whose degradation has been investigated include:

Hydrazine (diamide or diamine)	bp 113°C
Methylhydrazine (hydrazomethane or MMH)	bp 87°C
1,1-Dimethylhydrazine[27]	bp 63°C
1,2-Dimethylhydrazine dihydrochloride[28]	mp 167–169°C
1,1-Diethylhydrazine	bp 96–99°C
1,1-Diisopropylhydrazine	bp 41°C/16 mm Hg
1,1-Dibutylhydrazine	bp 87–90°C/21 mm Hg
N-Aminopyrrolidine hydrochloride	mp 117–119°C
N-Aminopiperidine (1-piperidinamine)	bp 146°C/730 mm Hg
N-Aminomorpholine (4-aminomorpholine)	bp 168°C
N,N'-Diaminopiperazine hydrochloride	
1-Methyl-1-phenylhydrazine	bp 54–55°C/0.3 mm Hg
1,2-Diphenylhydrazine (hydrazobenzene)	mp 123–126°C
Phenylhydrazine (hydrazinobenzene)	bp 238–241°C
p-Tolylhydrazine hydrochloride	mp >200°C
1,5-Diphenylcarbazide[29]	mp 175–177°C
Procarbazine hydrochloride[30]	mp 223–226°C
Isoniazid[31]	mp 171–173°C
Iproniazid phosphate[32]	mp 180–182°C

Hydrazines find wide use in synthetic organic chemistry, in cancer research laboratories, and in industry (for example, as rocket fuel). Some hydrazines, for example, procarbazine, isoniazid, and iproniazid, are used as drugs.

Principles of Destruction

Hydrazines may be degraded by reduction with nickel-aluminum (Ni-Al) alloy in potassium hydroxide (KOH) solution[33] or by oxidation with potassium permanganate in sulfuric acid ($KMnO_4$ in H_2SO_4) or sodium hypochlorite.[26] The products of the reductive reactions are the corresponding amines (found in 73–100% yield) with ring reduction also observed in the case of isoniazid and iproniazid;[34] the products of the oxidative reactions are not known. It has been shown, however, that oxidation of hydrazines tends to produce the highly carcinogenic nitrosamines as byproducts, particularly when 1,1-dimethylhydrazine is the substrate, as well as producing mutagenic residues.[35] The oxidative methods are only recommended for treating spills and cleaning glassware when the heterogeneous nature of the Ni-Al alloy reductive method renders it inapplicable. All of these methods produced >99% destruction of the hydrazines. Reduction of 1,2-diphenylhydrazine under acidic conditions might give rise to the carcinogen benzidine but it was found that reduction with Ni-Al alloy under basic conditions did not give rise to any detectable amount of benzidine (detection limit = 1% of the theoretical amount).[33]

Destruction Procedures[26,33]

Destruction of Bulk Quantities of Hydrazines

Dissolve the hydrazine in H_2O so that the concentration does not exceed 10 mg/mL. If the hydrazine is not sufficiently soluble in H_2O, use methanol instead. Add an equal volume of KOH solution (1 M) and stir the mixture magnetically. For every 100 mL of this solution add 5 g of Ni-Al alloy at such a rate that excessive frothing does not occur. The reaction can be quite exothermic. Do it in a reaction vessel whose volume is at least three times that of the final reaction mixture. Cover the reaction mixture, stir for 24 h, then filter it through a pad of Celite. Allow the spent nickel to dry on a metal tray for 24 h (away from flammable solvents) and discard it with the solid waste. Check the filtrate for completeness of destruction, neutralize, and discard it.

Destruction of Hydrazines in Aqueous Solution

Dilute the mixture with H_2O, if necessary, so that the concentration does not exceed 10 g/L. Add an equal volume of KOH solution (1 M) and stir the mixture magnetically. For every 100 mL of this solution add 5 g of Ni-Al alloy at such a rate that excessive frothing does not occur. The reaction can be quite exothermic. Do it in a reaction vessel whose volume is at least three times that of the final reaction mixture. Cover the reaction mixture, stir for 24 h, then filter it through a pad of Celite. Allow the spent nickel to dry on a metal tray for 24 h (away from flammable solvents) and discard it with the solid waste. Check the filtrate for completeness of destruction, neutralize, and discard it.

Destruction of Hydrazines in Organic Solvents Not Miscible with Water (For Example, Dichloromethane)

Dilute the solution, if necessary, so that the hydrazine concentration does not exceed 5 g/L. Stir the reaction mixture and add one volume of KOH solution (2 M) and three volumes of methanol [i.e., dichloromethane: H_2O: methanol 1:1:3]. For every liter of this solution add 100 g of Ni-Al alloy at such a rate that excessive frothing does not occur. The reaction can be quite exothermic. Do it in a reaction vessel whose volume is at least three times that of the final reaction mixture. Cover the reaction mixture, stir for 24 h, then filter it through a pad of Celite. Allow the spent nickel to dry on a metal tray for 24 h (away from flammable solvents) and discard it with the solid waste. Check the filtrate for completeness of destruction, neutralize, and discard it.

Destruction of Hydrazines in Alcohols

Dilute the solution, if necessary, so that the hydrazine concentration does not exceed 10 g/L. Add an equal volume of KOH solution (1 M) and stir the mixture magnetically. For every 100 mL of this solution add 5 g of Ni-Al alloy at such a rate that excessive frothing does not occur. The reaction can be quite exothermic. Do it in a reaction vessel whose volume is at least three times that of the final reaction mixture. Cover the reaction mixture, stir for 24 h, then filter it through a pad of Celite. Allow the spent nickel to dry on a metal tray for 24 h (away from flammable solvents) and discard it with the solid waste. Check the filtrate for completeness of destruction, neutralize, and discard it.

Destruction of Hydrazines in Dimethyl Sulfoxide (DMSO)

Dilute the mixture, if necessary, with H_2O or methanol so that the hydrazine concentration does not exceed 10 g/L. Add an equal volume of

KOH solution (1 M) and stir the mixture magnetically. For every 100 mL of this solution add 5 g of Ni-Al alloy at such a rate that excessive frothing does not occur. The reaction can be quite exothermic. Do it in a reaction vessel whose volume is at least three times that of the final reaction mixture. Cover the reaction mixture, stir for 24 h, then filter it through a pad of Celite. Allow the spent nickel to dry on a metal tray for 24 h (away from flammable solvents) and discard it with the solid waste. Check the filtrate for completeness of destruction, neutralize, and discard it.

Destruction of Hydrazines in Olive Oil and Mineral Oil

For each volume of solution add two volumes of petroleum ether (boiling range 90–100°C) and extract this mixture with one volume of 1 M hydrochloric acid (HCl). Extract the organic layer twice more with 1 M HCl and combine the extracts. Dilute the extracts, if necessary, so that the concentration of hydrazine does not exceed 10 mg/mL. Add an equal volume of KOH solution (2 M) and stir the mixture magnetically. For every 100 mL of this solution add 5 g of Ni-Al alloy at such a rate that excessive frothing does not occur. The reaction can be quite exothermic. Do it in a reaction vessel whose volume is at least three times that of the final reaction mixture. Cover the reaction mixture, stir for 24 h, then filter it through a pad of Celite. Allow the spent nickel to dry on a metal tray for 24 h (away from flammable solvents) and discard it with the solid waste. Check the filtrate for completeness of destruction, neutralize, and discard it.

Destruction of Hydrazines in Agar Gel

Add the contents of the agar plate (~ 17 g) to 75 mL of KOH solution (1 M) and stir and warm the mixture until the agar dissolves. Note that many hydrazines are liable to volatilize under these conditions. When cool add 2 g of Ni-Al alloy slowly enough so that excessive frothing does not occur. (If >100 mg of hydrazine is present, increase the weight of alloy and volume of KOH solution proportionately.) Stir the mixture for 24 h, then filter it through a pad of Celite. Allow the spent nickel to dry on a metal tray away from flammable solvents for 24 h, then discard it with the solid waste. Check the filtrate for completeness of destruction, neutralize, and discard it.

Dealing With Spills of Hydrazines

A. Add 3 M H_2SO_4 to the spill and remove as much of the spill as possible with an absorbent. Cover the residue with a 47.4 g/L solution of $KMnO_4$ in

3 M H_2SO_4. Leave this mixture overnight, then decolorize with ascorbic acid or sodium ascorbate, neutralize, and remove. If possible, check for completeness of decontamination by taking wipe samples and analyze these samples. Note that this procedure may damage painted surfaces or Formica. Place the absorbent material in an appropriate solvent and decontaminate it.

Note. This procedure frequently generates nitrosamines and results in mutagenic residues. It should only be used with caution and in particular it should not be used to treat spills of 1,1-dimethylhydrazine as the procedure could generate large quantities of N-nitrosodimethylamine.[35]

B. Remove as much of the spill as possible with an absorbent and then cover the residue with 5.25% sodium hypochlorite solution and leave overnight. Use fresh sodium hypochlorite solution (see below for assay procedure).

Note. This procedure frequently generates nitrosamines and results in mutagenic residues. It should only be used with caution.[35]

Decontamination of Equipment Contaminated with Hydrazines

A. Fill the equipment with a 47.4 g/L solution of $KMnO_4$ in 3 M H_2SO_4 (or cover the contaminated surface with this solution). After leaving overnight the solution should still be purple. If not, replace with fresh solution, which should stay purple for at least 1 h. At the end of the reaction, decolorize with sodium ascorbate or ascorbic acid, neutralize, and discard the solution, then clean the equipment.

Note. This procedure frequently generates nitrosamines and results in mutagenic residues. It should only be used with caution and in particular it should not be used to treat equipment contaminated with 1,1-dimethylhydrazine as the procedure could generate large quantities of N-nitrosodimethylamine.[35]

B. Immerse the equipment in a 5.25% sodium hypochlorite solution and leave overnight. Use fresh sodium hypochlorite solution (see below for assay procedures).

Note. This procedure frequently generates nitrosamines and results in mutagenic residues. It should only be used with caution.[35]

Assay of Sodium Hypochlorite Solution

Sodium hypochlorite solutions tend to deteriorate with time so they should be periodically checked for the amount of active chlorine they contain.

Pipette 10 mL of sodium hypochlorite solution into a 100-mL volumetric flask and fill to the mark with distilled H_2O. Pipette 10 mL of this solution into a conical flask containing 50 mL of distilled H_2O, 1 g of potassium iodide, and 12.5 mL of 2 M acetic acid. Titrate this solution against 0.1 N sodium thiosulfate solution using starch as an indicator. Each 1 mL of the sodium thiosulfate solution corresponds to 3.545 mg of active chlorine. The sodium hypochlorite solution used in these degradation reactions should contain 25–30 g of active chlorine/L.

Analytical Procedures

Many procedures have been published for the analysis of hydrazines. The more volatile hydrazines are conveniently analyzed by gas chromatography (GC).[33] For analysis by GC a 1.8 m × 2-mm i.d. packed column can be used. For 1,1-dimethylhydrazine (80°C), 1,1-diethylhydrazine (80°C), 1,1-diisopropylhydrazine (80°C), N-aminopyrrolidine (80°C), 1,1-dibutylhydrazine (90°C), N-aminopiperidine (100°C), and N-aminomorpholine (130°C) the packing is 10% Carbowax 20 M + 2% KOH on 80/100 Chromosorb W AW, for N,N'-diaminopiperazine (100°C), phenylhydrazine (130°C), and p-tolylhydrazine (130°C) the packing is 2% Carbowax 20 M + 1% KOH on 80/100 Supelcoport, for 1,2-dimethylhydrazine (100°C) the packing is 28% Pennwalt 223 + 4% KOH on 80/100 Gas Chrom R, and for 1-methyl-1-phenylhydrazine (110°C) and 1,2-diphenylhydrazine (150°C) the packing is 3% SP 2401-DB on 100/120 Supelcoport. The oven temperatures shown above in parentheses are only a guide; the exact conditions would have to be determined experimentally. The HPLC conditions for procarbazine and iproniazid are a 250 × 4.6-mm i.d. column of Microsorb C_8 with a mobile phase of methanol: 0.04% ammonium dihydrogen phosphate buffer 50:50 flowing at 1 mL/min. A UV detector set at 254 nm was used. For isoniazid the same equipment was used but the methanol : buffer ratio was 10:90.[34]

Mutagenicity Assays

A number of hydrazines have been found to be mutagenic. The reaction mixtures obtained from the Ni-Al alloy reactions were not tested for mutagenicity but similar reaction mixtures from the degradation of nitrosamines (hydrazines have been shown to be intermediates in this degradation[36]) were found not to be mutagenic.[37]

Related Compounds

The procedures listed above should be generally applicable to hydrazines.

References

1. International Agency for Research on Cancer. *IARC Monographs on the Evaluation of Carcinogenic Risk of Chemicals to Man*. Volume 4, *Some Aromatic Amines, Hydrazine and Related Substances,* N-*Nitroso Compounds and Miscellaneous Alkylating Agents*; International Agency for Research on Cancer: Lyon, 1974; pp 127–136.

2. International Agency for Research on Cancer. *IARC Monographs on the Evaluation of the Carcinogenic Risk of Chemicals to Humans, Supplement No. 4, Chemicals, Industrial Processes and Industries Associated with Cancer in Humans. IARC Monographs, Volumes 1 to 29*; International Agency for Research on Cancer: Lyon, 1982; pp 136–138.

3. International Agency for Research on Cancer. *IARC Monographs on the Evaluation of the Carcinogenic Risk of Chemicals to Humans, Supplement No. 7, Overall Evaluations of Carcinogenicity: An Updating of* IARC Monographs *Volumes 1 to 42*; International Agency for Research on Cancer: Lyon, 1987; pp 223–224.

4. Reference 1, pp 137–143.

5. Reference 1, pp 145–152.

6. Toth, B.; Shimizu, H. Methylhydrazine tumorigenesis in Syrian golden hamsters and the morphology of malignant histiocytomas. *Cancer Res.* **1973**, *33*, 2744–2753.

7. Sax, N.I; Lewis, R.J., Sr. *Dangerous Properties of Industrial Materials*, 7th ed.; Van Nostrand-Reinhold: New York, 1989; p. 1898.

8. Reference 7, pp 2745–2746.

9. Kelly, M.G.; O'Gara, R.W.; Yancey, S.T.; Botkin, C. Induction of tumors in rats with procarbazine hydrochloride. *J. Natl. Cancer Inst.* **1968**, *40*, 1027–1051.

10. Kelly, M.G.; O'Gara, R.W.; Yancey, S.T.; Gadekar, K.; Botkin, C.; Oliverio, V.T. Comparative carcinogenicity of *N*-isopropyl-α-(2-methylhydrazino)-*p*-toluamide.HCl (procarbazine hydrochloride), its degradation products, other hydrazines, and isonicotinic acid hydrazide. *J. Natl. Cancer Inst.* **1969**, *42*, 337–344.

11. International Agency for Research on Cancer. *IARC Monographs on the Evaluation of the Carcinogenic Risk of Chemicals to Humans*. Volume 26, *Some Antineoplastic and Immunosuppressive Agents*; International Agency for Research on Cancer: Lyon, 1981; pp 311-339.

12. Reference 2, pp 220–221.

13. Reference 3, pp 327–328.

14. Reference 2, pp 146–148.

15. Reference 3, pp 227–228.

16. Reference 1, pp 159–172.

17. Reference 7, pp 2876–2877.

18. Reference 7, pp 2320–2321.

19. Reference 7, p. 2036.

20. Reference 7, pp 2034–2035.

21. Reference 7, p. 2372.

22. Fieser, L.F.; Fieser, M. *Reagents for Organic Synthesis*; Wiley: New York, 1967; Vol. 1, p. 434.

23. Reference 7, pp 1894–1895.

24. Reference 7, pp 1396–1397.

25. Reference 7, pp 1397–1398.

26. Castegnaro, M.; Ellen, G.; Lafontaine, M.; van der Plas, H.C.; Sansone, E. B.; Tucker, S.P., Eds. *Laboratory Decontamination and Destruction of Carcinogens in Laboratory Wastes: Some Hydrazines*; International Agency for Research on Cancer: Lyon, 1983 (IARC Scientific Publications No. 54).

27. Other names are asymmetric dimethylhydrazine, *unsym*-dimethylhydrazine, *N,N*-dimethylhydrazine, Dimazine, and UDMH.

28. Other names are *N,N'*-dimethylhydrazine, *sym*-dimethylhydrazine, and SDMH. This compound is usually supplied as the dihydrochloride.

29. Another name is 2,2'-diphenylcarbonic dihydrazide.

30. Other names are *N*-(1-methylethyl)-4-((2-methylhydrazino)methyl)benzamide, ibenzmethyzin, *N*-4-isopropylcarbamoylbenzyl-*N'*-methylhydrazine, 2-(*p*-isopropylcarbamoylbenzyl)-1-methylhydrazine, *N*-isopropyl-α-(2-methylhydrazino)-*p*-toluamide, matulane, *N*-isopropyl-*p*-[(2-methylhydrazino)methyl]benzamide, *p*-(*N¹*-methylhydrazinomethyl)-*N*-isopropylbenzamide, 1-methyl-2-[(isopropylcarbamoyl)benzyl]hydrazine, natulan, and MIH. This compound is generally supplied and used as the hydrochloride.

31. Other names are 4-pyridinecarboxylic acid hydrazide, isonicotinyl hydrazide, isonicotinoyl hydrazide, isonicotinylhydrazine, isonicotinoylhydrazine, isonicotinic acid hydrazide, INH, rimitsid, Cedin(Aerosol), Isocid, Neoxin, Hidrasonil, Ertuban, Antimicina, Hyzyd, Isonex, Unicozyde, Zonazide, Hycozid, Niconyl, Isonicazide, Isonicid, Isonicotan, Tubazid, Tibizide, Isobicina, Isozide, Isonilex, Isonindon, Isotebezid, Nicetal, Nikozid, Nitadon, Nyscozid, Pelazid, Raumanon, Retozide, RU-EF-Tb, Tebecid, Tisiodrazida, Tizide, Isozyd, Sauterazid, Niplen, TB-Vis, Tekazin, Isidrina, Hydrazid, Nevin, Cotinazin, Dinacrin, Ditubin, Mybasan, Neoteben, Niadrin, Nicozide, Nydrazid, Nidaton, Nicizina, Nicotibina, Pycazide, Pyricidin, Isolyn, Pyrizidin, Rimifon, Robisellin, Isonizide, Neumandin, Isocotin, Tubicon, Tyvid, Tisin, Tibinide, Tubilysin, Tubomel, Tubeco, Atcotibine, Vazadrine, Vederon, Isdonidrin, and Zinadon.

32. Other names are 4-pyridinecarboxylic acid 2-(1-methylethyl)hydrazide, 1-isonicotinoyl-2-isopropylhydrazine, 1-isonicotinyl-2-isopropylhydrazine, *N*-isopropyl isonicotinhydrazide, Euphozid, Marsilid, and isonicotinic acid 2-isopropylhydrazide. This compound is frequently supplied as the phosphate salt.

33. Lunn, G.; Sansone, E.B.; Keefer, L.K. Reductive destruction of hydrazines as an approach to hazard control. *Environ. Sci. Technol.* **1983**, *17*, 240–243.

34. Lunn, G.; Sansone, E.B. Reductive destruction of dacarbazine, procarbazine hydrochloride, isoniazid, and iproniazid. *Am. J. Hosp. Pharm.* **1987**, *44*, 2519–2524.

35. Castegnaro, M.; Brouet, I.; Michelon, J.; Lunn, G.; Sansone, E.B. Oxidative destruction

of hydrazines produces *N*-nitrosamines and other mutagenic species. *Am. Ind. Hyg. Assoc. J.* **1986**, *47*, 360–364.

36. Lunn, G.; Sansone, E.B.; Keefer,L.K. Safe disposal of carcinogenic nitrosamines. *Carcinogenesis* **1983**, *4*, 315–319.

37. Lunn, G. Unpublished observations.

MERCAPTANS AND INORGANIC SULFIDES

CAUTION! Refer to safety considerations section on page 6 before starting any of these procedures.

Mercaptans are organic compounds of the general formula R-SH, where R is an alkyl or aryl group. The alkyl compounds are also known as thiols; for example, methyl mercaptan is also methanethiol and ethyl mercaptan is also ethanethiol. The aromatic compounds are also called thiophenols, for example, 2-chlorophenyl mercaptan is also called 2-chlorothiophenol. Methyl mercaptan[1] is a gas at room temperature (bp 6°C), ethyl mercaptan is a low boiling liquid (bp 35°C), and most other mercaptans are liquids [for example, thiophenol (benzenethiol) bp 169°C] although some solids are known. Most liquid and some solid mercaptans are volatile and possess an overpowering objectionable odor. Methyl mercaptan is toxic by inhalation.[2] Ethyl mercaptan[3] is a skin and eye irritant and has effects on the central nervous system.[4] Thiophenol is an eye irritant and can cause dermatitis, headaches, and dizziness.[5] Mercaptans should only be handled in a properly functioning chemical fume hood. They are used in synthetic organic chemistry.

151

Inorganic sulfides are solids which release hydrogen sulfide, an extremely toxic gas with an unpleasant odor,[6] on treatment with acid. Sodium sulfide can explode on rapid heating or percussion.[7] These compounds are widely used in chemical laboratories.

Principle of Destruction

Mercaptans can be oxidized by sodium or calcium hypochlorite to the corresponding sulfonic acid. These compounds are generally nonvolatile and odorless. In a similar fashion, inorganic sulfide can be oxidized to sulfate.

Destruction Procedures[8]

A. Before starting the reaction prepare an ice bath in case cooling of the reaction is required. Stir a 5.25% sodium hypochlorite solution (2.5 L) at room temperature in a 5-L flask and add 0.5 mol of liquid mercaptan dropwise. If the mercaptan is a solid, first dissolve it in tetrahydrofuran (THF). Take up traces of mercaptan in the dropping funnel or the reagent container in THF and add them to the reaction mixture. Add inorganic sulfides as aqueous solutions. The mercaptan is seen to dissolve and the temperature rises indicating that the reaction has started. If there is no reaction after 10% of the mercaptan has been added, warm the mixture to 50°C. Only add more mercaptan after it is clear that the reaction has started. Add the mercaptan at such a rate that the temperature stays at ~ 45–50°C; if the temperature rises above this it may be necessary to use the ice bath. When the addition is complete, stir the mixture for 2 h, then discard it.

Sodium hypochlorite solutions deteriorate with time so they should be used fresh or periodically checked for the amount of active chlorine they contain. Pipette 10 mL of sodium hypochlorite solution into a 100-mL volumetric flask and fill to the mark with distilled H_2O. Pipette 10 mL of this solution into a conical flask containing 50 mL of distilled H_2O, 1 g of potassium iodide, and 12.5 mL of 2 M acetic acid. Titrate this solution against 0.1 N sodium thiosulfate solution using starch as an indicator. Each 1 mL of the sodium thiosulfate solution corresponds to 3.545 mg of active chlorine. The sodium hypochlorite solution used in these degradation reactions should contain 25–30 g of active chlorine/L.

B. Before starting the reaction prepare an ice bath in case cooling of the reaction is required. Stir 210 g of calcium hypochlorite in 1 L of H_2O at room temperature in a 5-L flask and most will soon dissolve. Add 0.5 mol of liquid mercaptan dropwise. If the mercaptan is a solid, first dissolve it in THF. Take up traces of mercaptan in the dropping funnel or the reagent container in THF and add them to the reaction mixture. Add inorganic sulfides as aqueous solutions. The mercaptan is seen to dissolve and the temperature rises indicating that the reaction has started. If there is no reaction after 10% of the mercaptan has been added, warm the mixture to 50°C. Only add more mercaptan after it is clear that the reaction has started. Add the mercaptan at such a rate that the temperature stays at ~ 45–50°C; if the temperature rises above this it may be necessary to use the ice bath. When the addition is complete stir the mixture for 2 h, then discard it.

Related Compounds

Although they have not been tested, these procedures would probably be effective for treating compounds containing S-S bonds, for example, methyl disulfide or phenyl disulfide.

References

1. Other names are mercaptomethane, thiomethyl alcohol, and methyl sulfhydrate.

2. Sax, N.I; Lewis, R.J., Sr. *Dangerous Properties of Industrial Materials*, 7th ed.; Van Nostrand-Reinhold: New York, 1989; p. 2217.

3. Other names are mercaptoethane, thioethyl alcohol, and ethyl sulfhydrate.

4. Reference 2, p. 1581.

5. Reference 2, p. 373.

6. Reference 2, p. 1912.

7. Reference 2, p. 3096.

8. National Research Council, Committee on Hazardous Substances in the Laboratory. *Prudent Practices for Disposal of Chemicals from Laboratories;* National Academy Press: Washington, DC, 1983; pp 65 and 86.

METHOTREXATE

CAUTION! Refer to safety considerations section on page 6 before starting any of these procedures.

The degradation of a number of antineoplastic drugs, including methotrexate, was investigated by the International Agency for Research on Cancer (IARC).[1] Methotrexate (**I**)[2] is a solid (mp of the monohydrate 185–204°C) and is insoluble in water and organic solvents, but it is soluble in dilute acid or base.

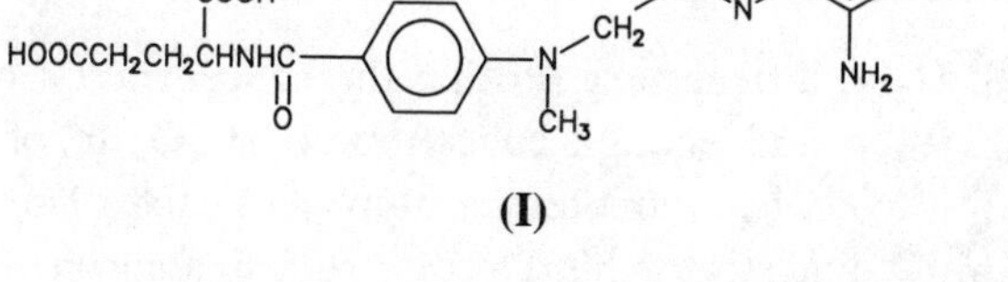

(I)

Methotrexate is not mutagenic to *Salmonella typhimurium* but it is mutagenic to *Streptococcus faecium*.[3] Methotrexate is a teratogen in hu-

155

mans and laboratory animals and can affect the blood, bone marrow, liver, and other organs.[4] It is used as an antineoplastic drug.

Principles of Destruction

Methotrexate is destroyed by oxidation with potassium permanganate in sulfuric acid ($KMnO_4$ in H_2SO_4), by oxidation with alkaline $KMnO_4$, or by oxidation with sodium hypochlorite. Destruction is >99.5% in all cases. The products of these reactions have not been established.

Destruction Procedures

Destruction of Bulk Quantities of Methotrexate

A. Dissolve in 3 M H_2SO_4 so that the concentration does not exceed 5 mg mL, then add 0.5 g of $KMnO_4$ for each 10 mL of solution and stir for 1 h. Decolorize with ascorbic acid, neutralize, check for completeness of destruction, and discard it.

B. Dissolve in 1 M sodium hydroxide (NaOH) solution so that the concentration does not exceed 1 mg/mL. For every 50 mL of this solution add 5.5 mL of a 1% (w/v) aqueous $KMnO_4$ solution and stir. The purple color should persist for at least 30 min; if it does not, add more $KMnO_4$ solution. Decolorize the reaction mixture with 0.1 M sodium bisulfite solution, neutralize, analyze for completeness of destruction, and discard it.

C. Dissolve in 1 M NaOH solution so that the concentration does not exceed 0.5 mg/mL. For every 100 mL of this solution add 10 mL of 5.25% sodium hypochlorite solution and stir for 30 min. Treat the reaction mixture with 0.1 M sodium bisulfite solution to remove excess oxidant, neutralize with 1 M hydrochloric acid (HCl) (**Caution!** Chlorine is evolved), analyze for completeness of destruction, and discard it. Use fresh sodium hypochlorite solution (see assay procedure below).

Destruction of Aqueous Solutions of Methotrexate

A. Dilute with H_2O, if necessary, so that the concentration does not exceed 5 mg/mL, then add enough concentrated H_2SO_4 to obtain a 3 M solution. Allow it to cool to room temperature. For each 10 mL of solution add 0.5 g of $KMnO_4$ and stir for 1 h. Decolorize with ascorbic acid, neutralize, check for completeness of destruction, and discard it.

B. Add at least an equal volume of 2 M NaOH solution, more if necessary, so that the concentration does not exceed 1 mg/mL. For every 50 mL of this solution add 5.5 mL of a 1% (w/v) aqueous $KMnO_4$ solution and stir. The purple color should persist for at least 30 min; if it does not, add more $KMnO_4$ solution. Decolorize the reaction mixture with 0.1 M sodium bisulfite solution, neutralize, analyze for completeness of destruction, and discard it.

C. For every 50 mg of methotrexate add 10 mL of 5.25% sodium hypochlorite solution and stir for 30 min. Then treat the reaction mixture with 0.1 M sodium bisulfite solution to remove excess oxidant, neutralize with 1 M HCl (**Caution!** Chlorine is evolved), analyze for completeness of destruction, and discard it. Use fresh sodium hypochlorite solution (see assay procedure below).

Destruction of Injectable Pharmaceutical Preparations of Methotrexate Containing 2–5% Glucose and 0.45% Saline

A. Dilute with H_2O so that the concentration does not exceed 2.5 mg/mL, then add enough concentrated H_2SO_4 to obtain a 3 M solution. Allow it to cool to room temperature. For each 10 mL of solution add 1 g of $KMnO_4$, in small portions to avoid frothing, and stir for 1 h. Decolorize with ascorbic acid, neutralize, check for completeness of destruction, and discard it.

B. Add at least an equal volume of 2 M NaOH solution, more if necessary, so that the concentration does not exceed 1 mg/mL. For every 50 mL of this solution add 5.5 mL of a 1% (w/v) aqueous $KMnO_4$ solution and stir. The purple color should persist for at least 30 min; if it does not, add more $KMnO_4$ solution. Decolorize the reaction mixture with 0.1 M sodium bisulfite solution, neutralize, analyze for completeness of destruction, and discard it.

C. For every 50 mg of methotrexate add 10 mL of 5.25% sodium hypochlorite solution and stir for 30 min. Treat the reaction mixture with 0.1 M sodium bisulfite solution to remove excess oxidant, neutralize with 1 M HCl (**Caution!** Chlorine is evolved), analyze for completeness of destruction, and discard it. Use fresh sodium hypochlorite solution (see assay procedure below).

Destruction of Solutions of Methotrexate in Volatile Organic Solvents

A. Remove the solvent under reduced pressure on a rotary evaporator and take up the residue in 3 M H_2SO_4 so that the concentration does not exceed 5 mg/mL. For each 10 mL of solution add 0.5 g of $KMnO_4$ and stir for 1 h. Decolorize with ascorbic acid, neutralize, check for completeness of destruction, and discard it.

B. Remove the solvent under reduced pressure on a rotary evaporator and take up the residue in 1 M NaOH solution so that the concentration does not exceed 1 mg/mL. For every 50 mL of this solution add 5.5 mL of a 1% (w/v) aqueous $KMnO_4$ solution and stir. The purple color should persist for at least 30 min; if it does not, add more $KMnO_4$ solution. Decolorize the reaction mixture with 0.1 M sodium bisulfite solution, neutralize, analyze for completeness of destruction, and discard it.

C. Remove the solvent under reduced pressure on a rotary evaporator and take up the residue in 1 M NaOH solution so that the concentration does not exceed 0.5 mg/mL. For every 100 mL of this solution add 10 mL of 5.25% sodium hypochlorite solution and stir for 30 min. Then treat the reaction mixture with 0.1 M sodium bisulfite solution to remove excess oxidant, neutralize with 1 M HCl (**Caution!** Chlorine is evolved), analyze for completeness of destruction, and discard it. Use fresh sodium hypochlorite solution (see assay procedure below).

Destruction of Dimethyl Sulfoxide (DMSO) or Dimethylformamide (DMF) Solutions of Methotrexate

Dilute with H_2O so that the concentration of DMSO or DMF does not exceed 20% and the concentration of the drug does not exceed 2.5 mg/mL, then add enough concentrated H_2SO_4 to obtain a 3 M solution. Allow it to cool to room temperature. For each 10 mL of solution add 1 g of $KMnO_4$ and stir for 1 h. Decolorize with ascorbic acid, neutralize, check for completeness of destruction, and discard it.

Decontamination of Glassware Contaminated with Methotrexate

A. Immerse the glassware in a 0.3 M solution of $KMnO_4$ in 3 M H_2SO_4 for 1 h, then clean by immersion in ascorbic acid solution.

B. Immerse the glassware in a mixture of 50 mL of 1 M NaOH solution and 5.5 mL of a 1% (w/v) aqueous $KMnO_4$ solution. After 30 min clean the glassware with 0.1 M sodium bisulfite solution.

C. Immerse the glassware in 5.25% sodium hypochlorite solution for 30 min. Use fresh sodium hypochlorite solution (see assay procedure below).

Decontamination of Spills of Methotrexate

A. Allow any organic solvent to evaporate and remove as much of the spill as possible by HEPA vacuuming (not sweeping), then rinse the area with 3

M H$_2$SO$_4$. Take up the rinse with absorbents and allow the rinse and absorbents to react with 0.3 M KMnO$_4$ solution in 3 M H$_2$SO$_4$ for 1 h. If the color fades, add more solution. Check for completeness of decontamination by using a wipe moistened with 0.1 M H$_2$SO$_4$. Analyze the wipe for the presence of the drug.

B. Allow any organic solvent to evaporate and remove as much of the spill as possible by HEPA vacuuming (not sweeping), then rinse the area with 1 M NaOH solution. Take up the rinse with absorbents and allow the rinse and absorbents to react with a mixture of 50 mL of 1 M NaOH solution and 5.5 mL of a 1% (w/v) aqueous KMnO$_4$ solution. Check for completeness of decontamination by using a wipe moistened with 0.1 M NaOH solution. Analyze the wipe for the presence of the drug.

C. Allow any organic solvent to evaporate and remove as much of the spill as possible by HEPA vacuuming (not sweeping), then rinse the area with 5.25% sodium hypochlorite solution then H$_2$O. Remove the rinses and discard them. Check for completeness of decontamination by using a wipe moistened with 0.1 M NaOH solution. Analyze the wipe for the presence of the drug. Use fresh sodium hypochlorite solution (see assay procedure below).

Assay of Sodium Hypochlorite Solution

Sodium hypochlorite solutions tend to deteriorate with time so they should be periodically checked for the amount of active chlorine they contain. Pipette 10 mL of sodium hypochlorite solution into a 100-mL volumetric flask and fill to the mark with distilled H$_2$O. Pipette 10 mL of this solution into a conical flask containing 50 mL of distilled H$_2$O, 1 g of potassium iodide, and 12.5 mL of 2 M acetic acid. Titrate this solution against 0.1 N sodium thiosulfate solution using starch as an indicator. Each 1 mL of the sodium thiosulfate solution corresponds to 3.545 mg of active chlorine. The sodium hypochlorite solution used in these degradation reactions should contain 25–30 g of active chlorine/L.

Analytical Procedures

Methotrexate can be analyzed by HPLC using a 25-cm reverse phase column and UV detection at 254 nm. Tetrabutylammonium phosphate (5 mM, adjusted to pH 3.5 with phosphoric acid) : methanol (55:45) flowing at 1.5 mL/min; 5 mM ammonium formate (adjusted to pH 5 with formic acid) :

methanol (60:40) flowing at 1 mL/min; and 0.1 M potassium phosphate, monobasic (KH_2PO_4) : methanol (80:20) flowing at 1 mL/min have been recommended as mobile phases.

Mutagenicity Assays

In the IARC study[1] tester strains TA100, TA1530, TA1535, and UTH8414 of *S. typhimurium* were used with and without mutagenic activation (not all strains were used for each procedure). The reaction mixtures were not mutagenic.

Related Compounds

Potassium permanganate in H_2SO_4 is a general oxidative method and should, in principle, be effective in destroying many drugs. However, any new application should be thoroughly validated both for complete destruction of the compound and for the production of nonmutagenic reaction mixtures.

References

1. Castegnaro, M.; Adams, J.; Armour, M-. A.; Barek, J.; Benvenuto, J.; Confalonieri, C.; Goff, U.; Ludeman, S.; Reed, D.; Sansone, E. B.; Telling, G., Eds. *Laboratory Decontamination and Destruction of Carcinogens in Laboratory Wastes: Some Antineoplastic Agents*; International Agency for Research on Cancer: Lyon, 1985 (IARC Scientific Publications No. 73).

2. Other names are N-[4-([(2,4-diamino-6-pteridinyl)methyl]methylamino)benzoyl]-L-glutamic acid, amethopterin, 4-amino-4-deoxy-N^{10}-methylpteroylglutamic acid, 4-amino-10-methylfolic acid, 4-amino-N^{10}-methylpteroylglutamic acid, N-bismethylpteroylglutamic acid, methopterin, methotextrate, MTX, Emtexate, A-Methopterin, and methylaminopterin.

3. International Agency for Research on Cancer. *IARC Monographs on the Evaluation of the Carcinogenic Risk of Chemicals to Humans, Supplement No. 4, Chemicals, Industrial Processes and Industries Associated with Cancer in Humans. IARC Monographs, Volumes 1 to 29*; International Agency for Research on Cancer: Lyon, 1982; pp 157–158.

4. Sax, N.I; Lewis, R.J., Sr. *Dangerous Properties of Industrial Materials*, 7th ed.; Van Nostrand-Reinhold: New York, 1989; pp 2221–2222.

2-METHYLAZIRIDINE

CAUTION! Refer to safety considerations section on page 6 before starting any of these procedures.

2-Methylaziridine (**I**) is a volatile liquid (bp 66–67°C) that has been reported to be a carcinogen in experimental animals.[1] It is a severe eye irritant, reacts vigorously with oxidizing materials, and can polymerize explosively.[2] It is used industrially as a chemical intermediate. It is miscible with H_2O and soluble in ethanol. Other names are 2-methylazacyclopropane, 2-methylethyleneimine, and 1,2-propyleneimine.

(I)

Principle of Destruction

2-Methylaziridine is reduced by nickel-aluminum alloy (Ni-Al) in potassium hydroxide (KOH) solution to give a mixture of isopropylamine and *n*-propylamine in an 85:15 ratio.[3] Destruction efficiency is >99.6%.

161

Destruction Procedure

Take up 2-methylaziridine (50 μL, 40.4 mg) in 10 mL of 1 M KOH solution and add 0.5 g of Ni-Al alloy. (If the reaction is done on a bigger scale, add the alloy in portions to avoid frothing.) Stir the reaction mixture for 18 h, then filter through a pad of Celite. Check the filtrate for completeness of destruction, neutralize, and discard it. Allow the spent nickel to dry on a metal tray for 24 h (away from flammable solvents) and discard it with the solid waste.[4]

Analytical Procedures[5]

Add 100 μL of the solution to be analyzed to 1 mL of a solution of 2 mL of acetic acid in 98 mL of 2-methoxyethanol. Swirl this mixture and add 1 mL of a solution of 5 g of 4-(4-nitrobenzyl)pyridine (4-NBP) in 100 mL of 2-methoxyethanol. Heat the solution at 100°C for 10 min, then cool in ice for 5 min. Add piperidine (0.5 mL) and 2-methoxyethanol (2 mL). Determine the violet color at 560 nm using a UV/Vis spectrophotometer.

Check the efficacy of the analytical procedure by adding a small quantity of 2-methylaziridine in 2-methoxyethanol solution to the solution to be analyzed after the acetic acid-2-methoxyethanol has been added but before the 4-NBP has been added. A positive response indicates that the analytical technique is satisfactory.

Using the analytical procedure described above with 10-mm disposable plastic cuvettes in a Gilford 240 UV/Vis spectrophotometer the limit of detection was 16 mg/L, but this can easily be reduced by using more than 100 μL of the solution to be analyzed.

The products were determined by gas chromatography using a 1.8 m × 2-mm i.d. packed column filled with 10% Carbowax 20 M + 2% KOH on 80/100 Chromosorb W AW. The oven temperature was 60°C, the injection temperature was 200°C, and the flame ionization detector operated at 300°C. Under these conditions, isopropylamine and n-propylamine had retention times of ~ 1.1 and 1.7 min, respectively. The GC conditions given above are only a guide and the exact conditions would have to be determined experimentally.

Mutagenicity Assays

The mutagenicity assays were carried out as described on page 4 using tester strains TA98, TA100, TA1530, and TA1535. The final reaction mixture (tested at a level corresponding to 0.4 mg of undegraded material) was not mutagenic although it was somewhat toxic to the cells. When it was diluted with an equal volume of pH 7 buffer it was not toxic and it was still not mutagenic. The pure compound was highly mutagenic, whereas the products, isopropylamine and *n*-propylamine, were not mutagenic.

Related Compounds

The procedure should be generally applicable to the destruction of other aziridines but it should be carefully checked to ensure that the compounds are completely degraded.

References

1. International Agency for Research on Cancer. *IARC Monographs on the Evaluation of the Carcinogenic Risk of Chemicals to Man.* Volume 9, *Some Aziridines,* N, S- *and* O-*Mustards and Selenium*; International Agency for Research on Cancer: Lyon, 1975; pp 61–65.

2. Sax, N.I; Lewis, R.J., Sr. *Dangerous Properties of Industrial Materials*, 7th ed.; Van Nostrand-Reinhold: New York, 1989; pp 2258–2259.

3. Lunn, G. Unpublished results.

4. Lunn, G.; Sansone, E.B.; Andrews, A.W.; Keefer, L.K. Decontamination and disposal of nitrosoureas and related *N*-nitroso compounds. *Cancer Res.* **1988**, *48*, 522–526.

5. This procedure was developed for determining dimethyl sulfate. See Lunn, G.; Sansone, E.B. Validation of techniques for the destruction of dimethyl sulfate. *Am. Ind. Hyg. Assoc. J.* **1985**, *46*, 111–114.

1-METHYL-4-PHENYL-1,2,3,6-TETRAHYDROPYRIDINE (MPTP)

CAUTION! Refer to safety considerations section on page 6 before starting any of these procedures.

1-Methyl-4-phenyl-1,2,3,6-tetrahydropyridine (MPTP) **(I)** is a colorless crystalline solid (mp 37–40°C), which produces the symptoms of Parkinson's disease in humans.[1] Other toxicological properties have not been established although it does not appear to be mutagenic.[2] It has a tendency to sublime and it should be handled with great care only in a properly functioning chemical fume hood. To avoid volatility problems, MPTP, which is a base, should only be handled in acidic solution. It is usually convenient to handle the compound as a nonvolatile salt. The preparation of the tartrate is described below. 1-Methyl-4-phenyl-1,2,3,6-tetrahydropyridine is used in the laboratory in research into Parkinson's disease.

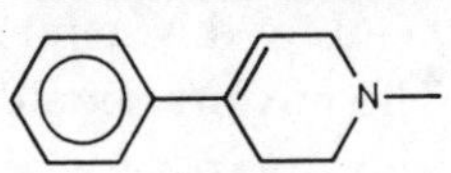

(I)

Principles of Destruction

1-Methyl-4-phenyl-1,2,3,6-tetrahydropyridine may be degraded by oxidation with potassium permanganate in 3 M sulfuric acid ($KMnO_4$ in H_2SO_4).[2] No detectable pyridinium oxidation product (1-methyl-4-phenylpyridinium ion) was formed.[2] Experiments using labeled MPTP showed that the products of the reaction were polar, H_2O-soluble species, probably carboxylic acids.[2] Reduction of MPTP to the physiologically inactive[3] 1-methyl-4-phenylpiperidine with nickel-aluminum alloy gave unsatisfactory results and it could not be recommended as a destruction procedure.[2]

Destruction Procedures

Destruction of Bulk Quantities

For every 25 mg of MPTP add 100 mL of 3 M H_2SO_4 and stir the mixture until it is homogeneous. For every 100 mL of solution add 4.7 g of $KMnO_4$ and stir the mixture overnight. The mixture should still be purple. If it is not, add more $KMnO_4$ until it stays purple for at least 1 h. Decolorize the mixture with sodium ascorbate or ascorbic acid, neutralize, analyze for completeness of destruction, and discard it.

Destruction of MPTP in Aqueous Solution

Dilute the solution with H_2O, if necessary, so that the concentration of MPTP does not exceed 0.25 mg/mL (or 0.4 mg/mL of MPTP tartrate), then add an equal volume of 6 M H_2SO_4. For every 100 mL of solution add 4.7 g of $KMnO_4$ and stir the mixture overnight. The mixture should still be purple. If it is not, add more $KMnO_4$ until it stays purple for at least 1 h. Decolorize the mixture with sodium ascorbate or ascorbic acid, neutralize, analyze for completeness of destruction, and discard it.

Destruction of MPTP in Ethanol, Methanol, Dimethyl Sulfoxide, and Acetone

Dilute the solution, if necessary, with the same solvent so that the concentration does not exceed 20 mg/mL. For every 1 mL of solution add 200 mL of 3 M H_2SO_4. For every 200 mL of this solution add 9.4 g of $KMnO_4$ and stir the mixture overnight. The mixture should still be purple. If it is not, add more $KMnO_4$ until it stays purple for at least 1 h. Decolorize the mixture with sodium ascorbate or ascorbic acid, neutralize, analyze for completeness of destruction, and discard it.

Destruction of MPTP in Acetonitrile:Aqueous Buffer HPLC Eluant

Note. This procedure applies **only** to acetonitrile : aqueous buffer HPLC eluant. It has not been tested for, and probably would not work with methanol : aqueous buffer mixtures.

To each volume of solution add two volumes of 6 M H_2SO_4 and for every 100 mL of this solution add 4.7 g of $KMnO_4$ and stir the mixture overnight. The mixture should still be purple. If it is not, add more $KMnO_4$ until it stays purple for at least 1 h. Decolorize the mixture with sodium ascorbate or ascorbic acid, neutralize, analyze for completeness of destruction, and discard it.

Preparation of MPTP Tartrate[4]

Add a solution of MPTP (5 g) in absolute ethanol (10 mL) to a solution of *d*-tartaric acid (4.5 g) in absolute ethanol (20 mL). Wash the original MPTP bottle with absolute ethanol (5 mL) and combine the washing with the other reagents. Warm the mixture until a clear solution forms and then allow it to cool to room temperature. Scratch to induce crystallization and filter the crystals, wash with cold ethanol and dry. Typical yields are 97%.

Analytical Procedures

Analysis may be performed by HPLC using a 5 μ C18 reverse phase column (25 cm). The mobile phase was acetonitrile : buffer (60:40) to which was added 0.1% triethylamine. The buffer consisted of 100 mM sodium acetate adjusted to pH 5.6 with acetic acid. The flow rate was 1 mL/min. A UV detector operating at 230 nm was used. The reaction mixtures were decolorized with sodium ascorbate, then neutralized by the addition of solid sodium bicarbonate before analysis. This mixture may be analyzed directly but problems were encountered when crystallization of the concentrated salt solution occurred. A better procedure was to place 250 μL of the decolorized and neutralized reaction mixture with 250 μL of acetonitrile in a small tube and stir for 5 min. The upper (acetonitrile) layer was then removed and analyzed by HPLC. Any MPTP that was present was extracted into the acetonitrile layer but the salts were left behind.

Mutagenicity Assays

The mutagenicity assays were carried out as described on page 4 using tester strains TA98, TA100, TA1535, TA1537, and TA1538. The final reaction mixtures [tested at a level corresponding to 62 μg of undegraded material (when bulk quantities were treated) per plate] were not mutagenic. The compounds MPTP, MPTP tartrate, 1-methyl-4-phenylpyridinium chloride, 1-methyl-4-phenylpiperidine hydrochloride, and 4-phenylpiperidine were also tested at concentrations of up to 1 mg/plate (2 mg for MPTP tartrate) and were not mutagenic.

Related Compounds

The initial oxidation product is probably the 1-methyl-4-phenylpyridinium ion. This compound can be detected using the same HPLC system used for MPTP. Since no trace of this ion was seen during the course of these oxidations,[2] this method can probably be used to degrade 1-methyl-4-phenylpyridinium chloride and related compounds containing the same cation.

References

1. Markey, S.P.; Schmuff, N.R. The pharmacology of the Parkinsonian syndrome producing neurotoxin MPTP (1-methyl-4-phenyl-1,2,3,6-tetrahydropyridine) and structurally related compounds. *Med. Res. Rev.* **1986**, *6*, 389–429.

2. Yang, S.C.; Markey, S.P.; Bankiewicz, K.S.; London, W.T.; Lunn, G. Recommended safe practices for using the neurotoxin MPTP in animal experiments. *Lab. Animal Sci.* **1988**, *38*, 563–567.

3. Cohen, G.; Mytilineou, C. Studies on the mechanism of action of 1-methyl-4-phenyl-1,2,3,6-tetrahydropyridine (MPTP). *Life Sci.* **1985**, *36*, 237–242.

4. Pitts, S.M.; Markey, S.P.; Murphy, D.L.; Weisz, A.; Lunn, G. Recommended practices for the safe handling of MPTP. In *MPTP - A Neurotoxin Producing a Parkinsonian Syndrome*; Markey, S.P., Castagnoli, Jr., N., Trevor, A., Kopin, I., Eds.; Academic Press: Orlando, 1986; pp 703–716.

MITOMYCIN C

CAUTION! Refer to safety considerations section on page 6 before starting any of these procedures.

Mitomycin C (**I**)[1] (mp >360°C) is a solid soluble in H_2O, methanol, acetone, butyl acetate, and cyclohexanone. Mitomycin C is carcinogenic in animals,[2] mutagenic,[3] and a teratogen.[4] It is used as an antineoplastic drug.

(I)

Principle of Destruction

Mitomycin C in urine can be destroyed by using sodium hypochlorite.[3]

Destruction Procedure

Destruction of Mitomycin C in Urine

For each 10 mL of urine add 1 mL of 5.25% sodium hypochlorite solution. Destruction is rapid. Destroy the excess sodium hypochlorite by adding 140 mg of sodium bisulfite. Neutralize the reaction mixture, check for completeness of destruction, and discard it. Use fresh sodium hypochlorite solution (see assay procedure below).

Assay of Sodium Hypochlorite Solution

Sodium hypochlorite solutions tend to deteriorate with time so they should be periodically checked for the amount of active chlorine they contain. Pipette 10 mL of sodium hypochlorite solution into a 100-mL volumetric flask and fill to the mark with distilled H_2O. Pipette 10 mL of this solution into a conical flask containing 50 mL of distilled H_2O, 1 g of potassium iodide, and 12.5 mL of 2 M acetic acid. Titrate this solution against 0.1 N sodium thiosulfate solution using starch as an indicator. Each 1 mL of the sodium thiosulfate solution corresponds to 3.545 mg of active chlorine. The sodium hypochlorite solution used in these degradation reactions should contain 25–30 g of active chlorine/L.

Analytical Procedures

Mitomycin C can be analyzed by HPLC using a 25-cm reverse phase column and UV detection at 254 nm. The mobile phase was methanol : H_2O 25:75 flowing at 1 mL/min.

Mutagenicity Assays

Tester strains TA100, TA102, UTH8413, and UTH8414 of *Salmonella typhimurium* were used without S9 activation.[3] Mitomycin C in urine was mutagenic to TA102. The final reaction mixtures were not mutagenic to any of the strains.

Related Compounds

Although oxidation with sodium hypochlorite works with Mitomycin C, the procedure should be thoroughly validated, including testing the final reaction mixtures for mutagenicity, before it is used for any other compound. For example, treatment of daunorubicin with sodium hypochlorite gave unacceptable results.[5]

References

1. Other names are [1aS-(1aα, 8β, 8aα, 8bα)]-6-amino-8-([(aminocarbonyl)oxy]methyl)-1,1a,2,8,8a,8b-hexahydro-8a-methoxy-5-methylazirino[2′,3′:3,4]pyrrolo[1,2-a]indole-4,7-dione, Ametycine, MMC, Mitocin-C, and Mutamycin.

2. International Agency for Research on Cancer. *IARC Monographs on the Evaluation of Carcinogenic Risk of Chemicals to Man*. Volume 10, *Some naturally occurring substances*; International Agency for Research on Cancer: Lyon, 1975; pp 171–179.

3. Monteith, D.K.; Connor, T.H.; Benvenuto, J.A.; Fairchild, E.J.; Theiss, J.C. Stability and inactivation of mutagenic drugs and their metabolites in the urine of patients administered antineoplastic therapy. *Environ. Mol. Mutagenesis* **1987**, *10*, 341-356.

4. Sax, N.I; Lewis, R.J., Sr. *Dangerous Properties of Industrial Materials*, 7th ed.; Van Nostrand-Reinhold: New York, 1989; pp 144–145.

5. Castegnaro, M.; Adams, J.; Armour, M-. A.; Barek, J.; Benvenuto, J.; Confalonieri, C.; Goff, U.; Ludeman, S.; Reed, D.; Sansone, E. B.; Telling, G., Eds. *Laboratory Decontamination and Destruction of Carcinogens in Laboratory Wastes: Some Antineoplastic Agents*; International Agency for Research on Cancer: Lyon, 1985 (IARC Scientific Publications No. 73).

4-NITROBIPHENYL

CAUTION! Refer to safety considerations section on page 6 before starting any of these procedures.

4-Nitrobiphenyl (4-NBP) is a crystalline solid (mp 113–114 °C), which is insoluble in H_2O, slightly soluble in alcohol, and soluble in organic solvents. Other names for 4-NBP are 4-nitro-1,1'-biphenyl, PNB, 4-nitrodiphenyl, and 4-phenylnitrobenzene. 4-Nitrobiphenyl causes cancer in laboratory animals.[1]

It is used as an intermediate in the chemical industry and in organic synthesis.

Principle of Destruction

4-Nitrobiphenyl is reduced by zinc in acid solution to 4-aminobiphenyl, which is then oxidized by potassium permanganate in sulfuric acid ($KMnO_4$ in H_2SO_4).[2] Less than 0.2% of 4-NBP or 4-ABP was left in the final reaction mixture.[3]

173

Destruction Procedure[2]

Dissolve bulk quantities of 4-NBP in glacial acetic acid so that the concentration does not exceed 1 mg/mL; dilute solutions in glacial acetic acid, if necessary, so that the concentration does not exceed 1 mg/mL, and evaporate organic solvents and take up the residue in glacial acetic acid so that the concentration does not exceed 1 mg/mL. To each of these solutions add an equal volume of 2 M H_2SO_4. Treat aqueous solutions by adding an equal volume of glacial acetic acid, then cautiously add, with stirring, 53 mL of concentrated H_2SO_4 per liter of solution. To each 20 mL of these acidified solutions add 165 mg of zinc powder with stirring. Stir the mixture overnight; then, for every 20 mL of solution, add 10 mL of 0.2 M $KMnO_4$ solution. Stir this mixture for 10 h, decolorize with ascorbic acid, neutralize, test for completeness of destruction, and discard it.

Analytical Procedures

The following HPLC analysis has been recommended.[2] A 250 × 4.6-mm i.d. reverse phase column was used and the mobile phase was MeCN : MeOH : buffer (10:30:20) flowing at 1.5 mL/min. The buffer was 1.5 mM in K_2HPO_4 and 1.5 mM in KH_2PO_4. If a variable wavelength UV detector is available, use 280 nm for 4-NBP and 275 nm for 4-ABP. Otherwise, a fixed wavelength detector operating at 254 nm should be satisfactory.

Mutagenicity Assays

The reaction mixtures were tested for mutagenicity using *Salmonella typhimurium* strains TA97, TA98, and TA100.[2] No mutagenic activity was seen.

References

1. International Agency for Research on Cancer. *IARC Monographs on the Evaluation of Carcinogenic Risk of Chemicals to Man*. Volume 4, *Some Aromatic Amines, Hydrazine and Related Substances, N-Nitroso Compounds and Miscellaneous Alkylating Agents*; International Agency for Research on Cancer: Lyon, 1974; pp 113–117.
2. Castegnaro, M.; Barek, J.; Dennis, J.; Ellen, G.; Klibanov, M.; Lafontaine, M.; Mitchum, R.; van Roosmalen, P.; Sansone, E.B.; Sternson, L.A.; Vahl, M., Eds. *Laboratory Decontamination and Destruction of Carcinogens in Laboratory Wastes: Some Aromatic Amines*

and 4-Nitrobiphenyl; International Agency for Research on Cancer: Lyon, 1985 (IARC Scientific Publications No. 64).

3. Barek, J.; Berka, A.; Müller, M.; Procházka, M.; Zima, J. Chemical destruction of 4-nitrobiphenyl in laboratory waste and its monitoring by differential pulse polarography and voltammetry and by high-performance liquid chromatography. *Coll. Czech. Chem. Commun.* **1986**, *51*, 1604–1608.

N-NITROSO COMPOUNDS: NITROSAMIDES

> **CAUTION!** Refer to safety considerations section on page 6 before starting any of these procedures.

Nitrosamides are compounds of the general form (**I**), where R is usually alkyl (for example, methyl or ethyl) and the X-N bond is labile. Compounds in which the X-N bond is not labile are termed nitrosamines. Since the chemistry of their decomposition is quite different, they are dealt with in a separate section. A particular problem with nitrosamides is base-induced decomposition to give diazoalkanes. For example, treatment of an ethanolic solution of *N*-methyl-*N*-nitroso-*p*-toluenesulfonamide with 1 *M* potassium hydroxide (KOH) solution generated diazomethane in 29% yield. Diazoalkanes are toxic, explosive, and carcinogenic and their generation should be avoided. All of the methods described below, with the exception of potassium permanganate in sulfuric acid ($KMnO_4$ in H_2SO_4), have been carefully checked for the generation of diazoalkanes; none were found. Because of the acid nature of the reaction medium, diazoalkane generation should not be a problem with $KMnO_4$ in H_2SO_4.

$$\begin{array}{c} R \\ \diagdown \\ \diagup \\ X \end{array} N{-}NO$$

(I)

The nitrosamides whose decomposition has been extensively studied are

MNTS	*N*-Methyl-*N*-nitroso-*p*-toluenesulfonamide[1]	mp 61–62°C
MNU	*N*-Methyl-*N*-nitrosourea	mp 126°C
ENU	*N*-Ethyl-*N*-nitrosourea	mp 104°C
MNUT	*N*-Methyl-*N*-nitrosourethane[2]	bp 62–64°C/ 12 mm Hg
ENUT	*N*-Ethyl-*N*-nitrosourethane[3]	bp 75°C/ 16 mm Hg
MNNG	*N*-Methyl-*N'*-nitro-*N*-nitrosoguanidine	mp 123°C
ENNG	*N*-Ethyl-*N'*-nitro-*N*-nitrosoguanidine	mp 118–120°C

Many nitrosamides cause cancer in laboratory animals and they should be regarded as potential human carcinogens. The International Agency for Research on Cancer has determined that MNU[4,5] and ENU[6,7] cause cancer in laboratory animals and should be regarded for practical purposes as carcinogenic to humans. The compounds MNNG,[8,9] MNUT,[10] and ENNG[11] cause cancer in laboratory animals while MNU,[12] MNNG,[13] MNUT,[14] ENU,[15] and ENUT[16] are teratogens. All the nitrosamides discussed in this section are mutagenic.[17] The compound MNNG will explode when heated or under impact,[13] MNUT explodes when heated,[14] and MNU can detonate[12] and has exploded on storage.[18] Nitrosamides should all be stored below -10°C. Some nitrosamides are solids and some are volatile liquids. They are generally soluble in polar organic solvents and only very sparingly soluble in H_2O.

As commercially supplied these nitrosamides may come in varying degrees of purity and this should be checked before commencing any experiments. For example, MNU and ENU are frequently supplied dampened with a little dilute acid to preserve their stability; purities in the 50–80% range are typical. **Unless absolutely necessary** these compounds should be used as supplied and purification should not be attempted because of the explosion hazard. Purity is best assessed by running a UV spectrum and comparing the observed absorbance with reported values.[19] If the purity is unacceptably low, obtain a fresh supply.

As far as possible these compounds should be handled in dilute solution using disposable equipment. Any reactions that may generate diazoalkanes should be done behind a safety shield using smooth glass apparatus with rubber stoppers and plastic tubing.

Nitrosamides are generally used in laboratories for the induction of tumors in experimental animals, although they also find some use in organic chemistry for the generation of diazoalkanes.

Principles of Destruction

Many methods have been investigated for the destruction of nitrosamides. The method of choice depends on the nitrosamide and the matrix in which it is found. The methods we will discuss are

Oxidation by potassium permanganate in sulfuric acid.[19] The products of the reaction have not been determined. Degradation efficiency was >99.5%.

Reaction with sulfamic acid in hydrochloric acid solution.[19] The strong hydrochloric acid (HCl) causes displacement of the nitroso group. The nitrosyl chloride formed reacts with the sulfamic acid to form nitrogen and H_2SO_4. This prevents any reformation of the nitrosamide. The products of the reaction are the corresponding amides produced by simple removal of the nitroso group. Degradation efficiency was >99.5%.

Reaction with iron filings in hydrochloric acid solution.[19] The strong HCl causes displacement of the nitroso group. The nitrosyl chloride formed is reduced by the iron filings in the acid to ammonia. This prevents any reformation of the nitrosamide. The products of the reaction are the corresponding amides produced by simple removal of the nitroso group except for MNNG and ENNG, where reductive removal of the nitro group causes the major products to be methylguanidine and ethylguanidine, respectively. Degradation efficiency was >99%.

Reaction with sodium bicarbonate solution.[20] This weak base causes a slow, base-mediated decomposition. The rate of reaction is sufficiently slow so that any diazoalkanes that are formed react with the solvent before escaping from the solution. The products of the reaction have not been definitely identified but they probably include methanol from MNU and MNUT, ethanol from ENU, MNUT, and ENUT, and cyanate from MNU and ENU.

Degradation efficiency was >99.99% for MNU, ENU, MNUT, and ENUT. The method is not suitable for MNNG, ENNG, or MNTS.

Reaction with sodium bicarbonate solution, then nickel-aluminum alloy and sodium carbonate solution, then potassium hydroxide solution.[21] The slow increase in pH of the solution produced by sequential addition of the bases causes a slow degradation of the nitrosamide. The degradation rate is sufficiently slow so that any diazoalkanes that are formed have time to react with the solvent before escaping from the solution. The products from this reaction have been discussed.[21] Degradation efficiency was >99.9%.

In all cases, using the procedures described below, destruction was complete, no diazoalkanes were detected, and the final reaction mixtures were not mutagenic.[17,19–21] Not all nitrosamides can be degraded by each procedure and these limitations are indicated at the start of each section. Provided that these restrictions are observed, the final reaction mixtures will not be mutagenic.

Destruction Procedures

Destruction of Bulk Quantities of Nitrosamides

A. Take up 1 g of the nitrosamide in 30 mL of methanol and add 30 mL of saturated aqueous sodium bicarbonate ($NaHCO_3$) solution. Stir the mixture at room temperature for 24 h, then add 30 mL of 1 M sodium carbonate (Na_2CO_3) solution and 10 g of nickel-aluminum (Ni-Al) alloy and stir for 24 h. At the end of this time add 30 mL of 1 M KOH solution (more if the solution has become too thick for easy stirring) and stir this mixture for 24 h, then filter it through a pad of Celite. Neutralize the filtrate, check for completeness of destruction, and discard it. Allow the solid to dry on a metal tray away from flammable solvents for 24 h, then discard it with the solid waste. For larger amounts of nitrosamide increase the quantities given here proportionately, but do this procedure **at least** on this scale. Dissolve small quantities of nitrosamides in **at least** 30 mL of methanol.

B. Dissolve the nitrosamide in 3 M H_2SO_4 so that the concentration does not exceed 5 g/L and add 47.4 g of $KMnO_4$ for each liter of solution. Stir the mixture at room temperature overnight, then decolorize with ascorbic acid, neutralize, check for completeness of destruction, and discard it.

C. For ENUT, MNUT, MNU, and ENU only. This method should **not** be used to degrade MNTS, MNNG, and ENNG. Take up 15 g of the nitrosamide in 1 L of ethanol and add 1 L of saturated $NaHCO_3$ solution. Stir the mixture for 24 h, neutralize, check for completeness of destruction, and discard it.

D. For MNTS, MNUT, MNU, ENU, MNNG, and ENNG only. This method should **not** be used to degrade ENUT. Take up 30 g of the nitrosamide in 1 L of methanol and slowly add 1 L of 6 M HCl with stirring. For each 2 L of solution add 70 g of sulfamic acid and stir the mixture for 24 h, neutralize, check for completeness of destruction, and discard it.

E. For MNTS, MNUT, MNU, and ENU only. This method should **not** be used to degrade ENUT, MNNG, and ENNG. Take up 30 g of the nitrosamide in 1 L of methanol and slowly add 1 L of 6 M HCl with stirring. For each 2 L of solution add 70 g of iron filings and stir the mixture for 24 h, neutralize, check for completeness of destruction, and discard it.

Destruction of Nitrosamides in Methanol

A. Dilute the solution, if necessary, with methanol so that the concentration of the nitrosamide does not exceed 1 g in 30 mL. For each 30 mL of this solution add 30 mL of saturated aqueous $NaHCO_3$ solution. Stir the mixture at room temperature for 24 h, then add 30 mL of 1 M Na_2CO_3 solution and 10 g of Ni-Al alloy and stir for 24 h. At the end of this time add 30 mL of 1 M KOH solution (more if the solution has become too thick for easy stirring) and stir this mixture for 24 h, then filter it through a pad of Celite. Neutralize the filtrate, check for completeness of destruction, and discard it. Allow the solid to dry on a metal tray away from flammable solvents for 24 h, then discard it with the solid waste. For larger amounts of nitrosamide increase the quantities given here proportionately, but do this procedure **at least** on this scale. Dissolve small quantities of nitrosamides in **at least** 30 mL of methanol.

B. Dilute the solution with H_2O so that the concentration of methanol does not exceed 20% and the concentration of the nitrosamide does not exceed 0.5%, then cautiously add 160 mL of concentrated H_2SO_4, with stirring, to each liter of solution. After cooling add 47.4 g of $KMnO_4$ for each liter of solution. Stir the mixture overnight at room temperature, decolorize with ascorbic acid, neutralize, check for completeness of destruction, and discard it.

C. For MNTS, MNUT, MNU, ENU, MNNG, and ENNG only. This method should **not** be used to degrade ENUT. Dilute the solution, if necessary, with methanol so that the concentration of the nitrosamide does not exceed 30 g/L. For each liter slowly add 1 L of 6 M HCl with stirring. For each 2 L of solution add 70 g of sulfamic acid, stir the mixture for 24 h, neutralize, check for completeness of destruction, and discard it.

D. For MNTS, MNUT, MNU, and ENU only. This method should **not** be used to degrade ENUT, MNNG, and ENNG. Dilute the solution, if necessary, with methanol so that the concentration of the nitrosamide does not exceed 30 g/L. For each liter slowly add 1 L of 6 M HCl with stirring. For each 2 L of solution add 70 g of iron filings, stir the mixture for 24 h, neutralize, check for completeness of destruction, and discard it.

Destruction of Nitrosamides in Ethanol

A. Dilute the solution,if necessary, with ethanol so that the concentration of the nitrosamide does not exceed 1 g in 30 mL. For each 30 mL of this solution add 30 mL of saturated aqueous $NaHCO_3$ solution. Stir the mixture at room temperature for 24 h, then add 30 mL of 1 M Na_2CO_3 solution and 10 g of Ni-Al alloy and stir for 24 h. At the end of this time add 30 mL of 1 M KOH solution (more if the solution has become too thick for easy stirring) and stir this mixture for 24 h, then filter it through a pad of Celite. Neutralize the filtrate, check for completeness of destruction, and discard it. Allow the solid to dry on a metal tray away from flammable solvents for 24 h, then discard it with the solid waste. For larger amounts of nitrosamide increase the quantities given here proportionately, but do this procedure **at least** on this scale. Dissolve small quantities of nitrosamides in **at least** 30 mL of ethanol.

B. Dilute the solution with H_2O so that the concentration of methanol does not exceed 20% and the concentration of the nitrosamide does not exceed 0.5%, then cautiously add 160 mL of concentrated H_2SO_4, with stirring, to each liter of solution. After cooling add 47.4 g of $KMnO_4$ for each liter of solution. Stir the mixture overnight at room temperature, decolorize with ascorbic acid, neutralize, check for completeness of destruction, and discard it.

C. For MNTS, MNU, ENU, MNNG, and ENNG only. This method should **not** be used to degrade MNUT and ENUT. Dilute the solution, if necessary, with ethanol so that the concentration of the nitrosamide does not exceed 30 g/L. For each liter slowly add 1 L of 6 M HCl with stirring. For each 2 L of solution add 70 g of sulfamic acid, stir the mixture for 24 h, neutralize, check for completeness of destruction, and discard it.

D. Dilute the solution, if necessary, with ethanol so that the concentration of the nitrosamide does not exceed 30 g/L. For each liter slowly add 1

L of 6 M HCl with stirring. For each 2 L of solution add 70 g of iron filings, stir the mixture for 24 h, neutralize, check for completeness of destruction, and discard it.

Destruction of Nitrosamides in Dimethyl Sulfoxide (DMSO)

A. Dilute the solution, if necessary, with DMSO so that the concentration of the nitrosamide does not exceed 1 g in 30 mL. For each 30 mL of this solution add 30 mL of saturated aqueous $NaHCO_3$ solution. Stir the mixture at room temperature for 24 h, then add 30 mL of 1 M Na_2CO_3 solution and 10 g of Ni-Al alloy and stir for 24 h. At the end of this time add 30 mL of 1 M KOH solution (more if the solution has become too thick for easy stirring) and stir this mixture for 24 h, then filter it through a pad of Celite. Neutralize the filtrate, check for completeness of destruction, and discard it. Allow the solid to dry on a metal tray away from flammable solvents for 24 h, then discard it with the solid waste. For larger amounts of nitrosamide increase the quantities given here proportionately, but do this procedure **at least** on this scale. Dissolve small quantities of nitrosamides in **at least** 30 mL of DMSO.

B. Dilute the solution with H_2O so that the concentration of DMSO does not exceed 20% and the concentration of the nitrosamide does not exceed 0.5%, then cautiously add 160 mL of concentrated H_2SO_4, with stirring, to each liter of solution. After cooling add 47.4 g of $KMnO_4$ for each liter of solution. Stir the mixture overnight at room temperature, decolorize with ascorbic acid, neutralize, check for completeness of destruction, and discard it.

C. For MNTS, MNUT, MNU, ENU, MNNG, and ENNG only. This method should **not** be used to degrade ENUT. Dilute the solution, if necessary, with DMSO so that the concentration of the nitrosamide does not exceed 30 g/L. For each liter slowly add 1 L of 6 M HCl with stirring. For each 2 L of solution add 70 g of sulfamic acid, stir the mixture for 24 h, neutralize, check for completeness of destruction, and discard it.

D. Dilute the solution, if necessary, with DMSO so that the concentration of the nitrosamide does not exceed 30 g/L. For each liter slowly add 1 L of 6 M HCl with stirring. For each 2 L of solution add 70 g of iron filings, stir the mixture for 24 h, neutralize, check for completeness of destruction, and discard it.

Destruction of Nitrosamides in Acetone

Dilute the solution, if necessary, with acetone so that the concentration of the nitrosamide does not exceed 1 g in 30 mL. For each 30 mL of this solution add 30 mL of saturated aqueous NaHCO₃ solution. Stir the mixture at room temperature for 24 h, then add 30 mL of 1 M Na₂CO₃ solution and 10 g of Ni-Al alloy and stir for 24 h. At the end of this time add 30 mL of 1 M KOH solution (more if the solution has become too thick for easy stirring) and stir this mixture for 24 h, then filter it through a pad of Celite. Neutralize the filtrate, check for completeness of destruction, and discard it. Allow the solid to dry on a metal tray away from flammable solvents for 24 h, then discard it with the solid waste. For larger amounts of nitrosamide increase the quantities given here proportionately, but do this procedure **at least** on this scale. Dissolve small quantities of nitrosamides in **at least** 30 mL of acetone.

Destruction of Nitrosamides in Water

A. For each 30 mL of solution add 30 mL of saturated aqueous NaHCO₃ solution. Stir the mixture at room temperature for 24 h, then add 30 mL of 1 M Na₂CO₃ solution and 10 g of Ni-Al alloy and stir for 24 h. At the end of this time add 30 mL of 1 M KOH solution (more if the solution has become too thick for easy stirring) and stir this mixture for 24 h, then filter it through a pad of Celite. Neutralize the filtrate, check for completeness of destruction, and discard it. Allow the solid to dry on a metal tray away from flammable solvents for 24 h, then discard it with the solid waste. For larger amounts of nitrosamide increase the quantities given here proportionately, but do this procedure **at least** on this scale. Dissolve small quantities of nitrosamides in **at least** 30 mL of H₂O.

B. Dilute the solution, if necessary, with H₂O so that the concentration of the nitrosamide does not exceed 0.5%, then cautiously add 160 mL of concentrated H₂SO₄, with stirring, to each liter of solution. After cooling add 47.4 g of KMnO₄ for each liter of solution. Stir the mixture overnight at room temperature, decolorize with ascorbic acid, neutralize, check for completeness of destruction, and discard it.

Destruction of Nitrosamides in Ethyl Acetate

Dilute the solution, if necessary, with ethyl acetate so that the concentration of the nitrosamide does not exceed 0.5%. Then, for each 30 mL of this solution, add 20 mL of dimethylformamide and 50 mL of a 0.3 M solution of KMnO₄ in H₂O. Cautiously add 16 mL of concentrated H₂SO₄, with stirring, and shake the mixture for ~ 1 min, then add 2.5 g of KMnO₄. Stir the mixture overnight at room temperature, decolorize with ascorbic acid, neutralize, check for completeness of destruction, and discard it.

Decontamination of Glassware

As far as possible drain the glassware of all the nitrosamide solution and treat this solution using one of the methods detailed above. Treat the drained glassware with one of the following methods.

A. Immerse the glassware overnight in a solution containing 3 M H$_2$SO$_4$ and 0.3 M KMnO$_4$. Clean the glassware by immersion in a 5% solution of ascorbic acid.

B. For glassware contaminated with ENUT, MNUT, MNU, and ENU only. This method should **not** be used to degrade MNTS, MNNG, and ENNG. Soak the glassware in **either** equal volumes of saturated NaHCO$_3$ solution and ethanol **or** the clear filtrate obtained when equal volumes of ethanol and saturated NaHCO$_3$ solution are mixed and filtered. After 2 h (MNU), 4 h (ENU), or 24 h (MNUT and ENUT), clean the glassware.

C. For glassware contaminated with MNTS, MNNG, and ENNG only. This method should **not** be used to degrade ENUT, MNUT, MNU, and ENU. Soak the glassware in a mixture of equal volumes of methanol and sulfamic acid in 2 M HCl (70 g/L). After 2 h (MNTS), 6 h (MNNG), or 24 h (ENNG), clean the glassware.

Treatment of Spills[17]

Remove as much material as possible by using absorbents or by HEPA vacuuming. Treat the material that is removed by one of the methods detailed above. Treat the contaminated area with one of the following methods.

A. For spills of ENUT, MNUT, MNU, and ENU only. This method should **not** be used for MNTS, MNNG, and ENNG. Add ethanol until all the nitrosamide appears to be dissolved, then add an equal volume of saturated NaHCO$_3$ solution. Alternatively, treat the area with the clear filtrate obtained when equal volumes of ethanol and saturated NaHCO$_3$ solution are mixed and filtered. After 2 h (MNU), 4 h (ENU), or 24 h (MNUT and ENUT), clean the area. At the end of the procedure check for completeness of destruction by using a wipe soaked in methanol and analyzing the wipe for the presence of the compound.

B. For spills of MNTS, MNNG, and ENNG only. This method should **not** be used for ENUT, MNUT, MNU, and ENU. Soak the area in methanol until the nitrosamide appears to be dissolved, then add an approximately equal volume of sulfamic acid in 2 M HCl (70 g/L). After 2 h

(MNTS), 6 h (MNNG), or 24 h (ENNG), clean the area. At the end of the procedure check for completeness of destruction by using a wipe soaked in methanol and analyzing the wipe for the presence of the compound.

C. Cover the area overnight with a solution containing 3 M H_2SO_4 and 0.3 M $KMnO_4$. Then treat the area with a 5% solution of ascorbic acid and, after 15 min, add solid Na_2CO_3 and clean the area. At the end of the procedure check for completeness of destruction by using a wipe soaked in methanol and analyzing the wipe for the presence of the compound.

Analytical Procedures[21]

The nitrosamides may be determined by HPLC using a UV detector operating at 254 nm and a 250 × 4.6-mm i.d. reverse phase column. The mobile phase is a mixture of methanol and 3.5 mM $(NH_4)H_2PO_4$ buffer flowing at 1 mL/min (except 2 mL/min for MNTS). The methanol : buffer ratios are 50:50 for MNTS, 40:60 for ENNG, MNUT, and ENUT, 12:88 for ENU and MNNG, and 7:93 for MNU.

To determine diazoalkanes, use smooth glass apparatus with plastic tubing and rubber stoppers behind a safety shield. Use a fresh reaction flask for each determination since scoured glass surfaces can cause decomposition of the diazoalkanes. Sweep the reaction with nitrogen flowing at 50–55 mL/min for a total of 2 h during and after the addition of the reagent to the nitrosamide. Allow the nitrogen to bubble through 75 mL of ether containing 2.5 mL of valeric acid. At the end of the reaction wash the ether solution twice with saturated aqueous $NaHCO_3$ solution and dry it over anhydrous magnesium sulfate. Add 100 μL of 1-butanol as an internal standard and analyze the mixture by gas chromatography using a 1.8 m × 2-mm i.d. glass column packed with 28% Pennwalt 223 + 4% KOH on 80/100 Gas Chrom R at 170°C. Methyl or ethyl valerate, produced by diazoalkane alkylation of the valeric acid, indicates the presence of diazomethane or diazoethane, respectively.

Toluene produced by the decomposition of MNTS interferes with this determination so butyric acid is substituted for valeric acid and 1-propanol is substituted for 1-butanol in the above procedure when analyzing reaction mixtures containing MNTS. Methyl butyrate is detected with the GC oven temperature set at 130°C. From time to time a positive control consisting of 1 g of MNTS in 30 mL of ethanol to which is added 30 mL of 1 M KOH solution should be used.

Reagents

Prepare saturated $NaHCO_3$ solution by occasionally shaking a mixture of $NaHCO_3$ and H_2O. Add solid $NaHCO_3$ until a precipitate persists. The supernatant is saturated $NaHCO_3$ solution.

Prepare 6 M HCl by cautiously adding concentrated HCl to an equal volume of H_2O. This is an exothermic reaction.

Mutagenicity Assays

The mutagenicity assays were carried out as described on page 4 using tester strains TA98, TA100, TA1530, and TA1535. The final reaction mixtures were not mutagenic.[17,20,21] The pure compounds were mutagenic,[17] whereas the products that could be identified were not.[20]

Related Compounds

In theory, the methods described above should be applicable to other nitrosamides, for example, N-methyl-N-nitrosoacetamide, but the chemistry of these compounds varies widely and the destruction procedure chosen **must** be fully validated before being employed on a routine basis. Validation includes determining that the nitrosamide is completely degraded, no diazoalkanes are generated, the final reaction mixtures are nonmutagenic and, if possible, the products are nontoxic. The destruction procedure should be validated for the nitrosamide dissolved in each solvent in which it is likely to be encountered. As an example of the problems that may arise, the mixed base/Ni-Al alloy method produced no mutagenic residues when applied to the seven nitrosamides whose physical characteristics are listed at the beginning of this section but did produce mutagens when applied to nitrosoureas containing β-chloroethyl groups, for example, 1,3-bis(2-chloroethyl)-1-nitrosourea (BCNU).[21]

References

1. Other names are N,4-dimethyl-N-nitrosobenzenesulfonamide, p-tolylsulfonyl methyl nitrosamine, p-tolylsulfonylmethylnitrosamide, and Diazald.
2. Another name is N-methyl-N-nitrosoethylcarbamate.
3. Another name is N-ethyl-N-nitrosocarbamic acid ethyl ester.

4. International Agency for Research on Cancer. *IARC Monographs on the Evaluation of Carcinogenic Risk of Chemicals to Man*. Volume 1; International Agency for Research on Cancer: Lyon, 1971; pp 125–134.

5. International Agency for Research on Cancer. *IARC Monographs on the Evaluation of the Carcinogenic Risk of Chemicals to Humans*. Volume 17, *Some* N-*Nitroso Compounds*; International Agency for Research on Cancer: Lyon, 1978; pp 227–255.

6. Reference 4, pp 135–140.

7. Reference 5, pp 191–215.

8. International Agency for Research on Cancer. *IARC Monographs on the Evaluation of the Carcinogenic Risk of Chemicals to Humans, Supplement No. 7, Overall Evaluations of Carcinogenicity: An Updating of* IARC Monographs *Volumes 1 to 42*; International Agency for Research on Cancer: Lyon, 1987; pp 248–250.

9. International Agency for Research on Cancer. *IARC Monographs on the Evaluation of Carcinogenic Risk of Chemicals to Man*. Volume 4, *Some Aromatic Amines, Hydrazine and Related Substances,* N-*Nitroso Compounds and Miscellaneous Alkylating Agents*; International Agency for Research on Cancer: Lyon, 1974; pp 183–195.

10. Reference 9, pp 211–220.

11. Nakamura, T.; Matsuyama, M.; Kishimoto, H. Tumors of the esophagus and duodenum induced in mice by oral administration of *N*-ethyl-*N'*-nitro-*N*-nitrosoguanidine. *J. Natl. Cancer Inst.* **1974**, *52*, 519–522.

12. Sax, N.I; Lewis, R.J., Sr. *Dangerous Properties of Industrial Materials*, 7th ed.; Van Nostrand-Reinhold: New York, 1989; pp 2358–2359.

13. Reference 12, p. 2349.

14. Reference 12, p. 2356.

15. Reference 12, pp 1666–1667.

16. Reference 12, pp 2552–2553.

17. Lunn, G.; Sansone, E.B. Dealing with spills of hazardous chemicals: Some nitrosamides. *Food Chem. Toxicol.* **1988**, *26*, 481–484.

18. Sparrow, A.H. Hazards of chemical carcinogens and mutagens. *Science* **1973**, *181*, 700–701.

19. Castegnaro, M.; Benard, M.; van Broekhoven, L. W.; Fine, D.; Massey, R.; Sansone, E.B.; Smith, P.L.R.; Spiegelhalder, B.; Stacchini, A.; Telling, G.; Vallon, J.J., Eds. *Laboratory Decontamination and Destruction of Carcinogens in Laboratory Wastes: Some* N-*Nitrosamides*; International Agency for Research on Cancer: Lyon, 1983 (IARC Scientific Publications No. 55).

20. Lunn, G. Unpublished results.

21. Lunn, G.; Sansone, E.B.; Andrews, A.W.; Keefer, L.K. Decontamination and disposal of nitrosoureas and related *N*-nitroso compounds. *Cancer Res.* **1988**, *48*, 522–526.

N-NITROSO COMPOUNDS: NITROSAMINES

CAUTION! Refer to safety considerations section on page 6 before starting any of these procedures.

Nitrosamines are compounds of the general form (**I**), where R and R' are usually aryl or alkyl, although other variants where R or R' are heteroatoms (for example, oxygen) are known. Compounds in which R or R' is a good leaving group are termed nitrosamides. Since the chemistry of their decomposition is quite different they are dealt with in a separate section.

$$\begin{array}{c} R \\ \diagdown \\ R' \diagup \end{array} N\!-\!NO$$

(I)

Many of the lower molecular weight nitrosamines are volatile liquids and they should all be regarded as carcinogenic in humans. For example, NDMA (*N*-methyl-*N*-nitrosomethanamine, *N*-nitrosodimethylamine, or dimethylnitrosamine) is a yellow-green H_2O-soluble liquid (bp 149–150°C), which has been shown to cause cancer in mice at the part per billion level in drinking H_2O. Nitrosamines are soluble in organic solvents and to varying degrees in H_2O.

Other commonly used nitrosamines whose degradation has been investigated include:

NDEA	*N*-Ethyl-*N*-nitrosoethanamine, *N*-nitrosodiethylamine, or diethylnitrosamine bp 177°C; soluble 106 mg/mL in H_2O
NDPA	*N*-Nitrosodipropylamine or dipropylnitrosamine bp 81°C/5 mm Hg; soluble 9.8 mg/mL in H_2O
NDiPA	*N*-Nitrosodiisopropylamine or diisopropylnitrosamine mp 48°C; slightly soluble in H_2O
NDBA	*N*-Nitrosodibutylamine[1] bp 116°C/14 mm Hg; soluble 1.2 mg/mL in H_2O
NPYR	*N*-Nitrosopyrrolidine or NO-PYR bp 98°C/12 mm Hg; miscible with H_2O
NPIP	*N*-Nitrosopiperidine bp 100°C/14 mm Hg; soluble 77 mg/mL in H_2O
NMOR	*N*-Nitrosomorpholine or 4-nitrosomorpholine bp 96°C/6 mm Hg; miscible with H_2O
di-NPZ	*N,N'*-Dinitrosopiperazine mp 160°C; soluble 5.7 mg/mL in H_2O
NMA	*N*-Nitroso-*N*-methylaniline[2] mp 15°C, bp 121°C/13 mm Hg; insoluble in H_2O
NDPhA	*N*-Nitrosodiphenylamine mp 66.5°C

The International Agency for Research on Cancer has determined that NDMA,[3,4] NDEA,[5,6] NDPA,[7] NDBA,[8,9] NPYR,[10] NPIP,[11] and NMOR[12] cause cancer in laboratory animals and should be regarded for practical purposes as carcinogenic to humans. The compounds NDiPA,[13] di-NPZ,[14] NMA,[15] and NDPhA[16] cause cancer in laboratory animals. *N*-Nitrosodimethylamine causes fatal liver disease and a number of systemic effects upon ingestion.[17] *N*-Nitrosodiphenylamine is an eye irritant and can react vigorously with oxidizing materials.[18] The compounds NDMA,[17] NDEA,[19] NDPA,[20] NDBA,[21] di-NPZ,[22] and NMA[23] are teratogens.

Nitrosamines are generally used in laboratories for the induction of tumors in experimental animals although they find some use as intermediates in organic chemistry, for example, for the preparation of hydrazines or in the α-alkylation of amines. They are also found as unwanted byproducts of industrial processes, for example, in the rubber industry.

Principles of Destruction

Nitrosamines may be reduced to the corresponding amine by using nickel-aluminum (Ni-Al) alloy in dilute base.[24] The nitrosamines were completely degraded (99.9%) and only the amines (RR'NH) were found in the final reaction mixtures. No traces (generally <0.1%) of the corresponding, possibly carcinogenic, hydrazines (RR'NNH$_2$) were found in the final reaction mixtures.

Nitrosamines may be oxidized by potassium permanganate in 3 M sulfuric acid (KMnO$_4$ in H$_2$SO$_4$).[25] The nitrosamines were completely destroyed (>99.5%). The products of this reaction have not been determined.

Nitrosamines may be destroyed by using hydrogen bromide (HBr) in glacial acetic acid.[25] The nitrosamines were completely destroyed (>99%) and the products were presumably the corresponding amines.

All of these procedures were validated by an international collaborative study.[25]

Destruction Procedures[24,25]

Destruction of Bulk Quantities of Nitrosamines

A. Dissolve the nitrosamine in H$_2$O so that the concentration does not exceed 10 mg/mL. If the nitrosamine is not sufficiently soluble in H$_2$O, use methanol instead. Add an equal volume of potassium hydroxide (KOH) solution (1 M) and stir the mixture magnetically. For every 100 mL of this solution add 5 g of Ni-Al alloy at such a rate that excessive frothing does not occur. The reaction can be quite exothermic. Do it in a reaction vessel whose volume is at least three times that of the final reaction mixture. Cover the reaction mixture and stir for 24 h, then filter it through a pad of Celite. Neutralize the filtrate, check for completeness of destruction, and discard it. Allow the spent nickel to dry on a metal tray for 24 h (away from flammable solvents) and discard it.

B. Dissolve the nitrosamine in H$_2$SO$_4$ (3 M) so that the nitrosamine concentration does not exceed 6 mg/L and then add 47.4 g of KMnO$_4$ to each liter. Stir the reaction mixture overnight. If the reaction mixture is no longer purple, add more $KMnO_4$ until the purple color is maintained for at least 1 h. Decolorize the reaction mixture with ascorbic acid or sodium ascorbate, neutralize, check for completeness of destruction, and discard it.

Destruction of Nitrosamines in Aqueous Solution

A. Dilute the mixture, if necessary, with H_2O so that the concentration does not exceed 10 mg/mL. Add an equal volume of KOH solution (1 M) and stir the mixture magnetically. For every 100 mL of this solution add 5 g of Ni-Al alloy at such a rate that excessive frothing does not occur. The reaction can be quite exothermic. Do it in a reaction vessel whose volume is at least three times that of the final reaction mixture. Cover the reaction mixture and stir for 24 h, then filter it through a pad of Celite. Neutralize the filtrate, check for completeness of destruction, and discard it. Allow the spent nickel to dry on a metal tray for 24 h (away from flammable solvents) and discard it with the solid waste.

B. Stir the aqueous solution and slowly add concentrated H_2SO_4 **to the aqueous solution** so that the H_2SO_4 concentration is 3 M. If necessary, use a cold water bath. Add 3 M H_2SO_4, if necessary, so that the nitrosamine concentration does not exceed 6 mg/L. Add 47.4 g of $KMnO_4$ to each liter of solution and stir the mixture overnight. If the reaction mixture is no longer purple, add more $KMnO_4$ until the purple color is maintained for at least 1 h. Decolorize the reaction mixture with ascorbic acid or sodium ascorbate, neutralize, check for completeness of destruction, and discard it.

Destruction of Nitrosamines in Aprotic Organic Solvents (For Example, Dichloromethane)

A. Dilute the solution, if necessary, so that the nitrosamine concentration does not exceed 10 mg/mL. Stir the reaction mixture and add one volume of KOH solution (2 M) and three volumes of methanol (i.e., dichloromethane: H_2O: methanol 1:1:3). For every 100 mL of this solution add 5 g of Ni-Al alloy at such a rate that excessive frothing does not occur. The reaction can be quite exothermic. Do it in a reaction vessel whose volume is at least three times that of the final reaction mixture. Cover the reaction mixture, stir for 24 h, then filter it through a pad of Celite. Neutralize the filtrate, check for completeness of destruction, and discard it. Allow the spent nickel to dry on a metal tray for 24 h (away from flammable solvents) and discard it with the solid waste.

B. Dry the solution, if necessary, with sodium sulfate. If necessary, dilute the mixture with more solvent so that the nitrosamine concentration does not exceed 1 mg/mL. For every volume of nitrosamine solution add 10 volumes of a 3% solution of HBr in glacial acetic acid (obtained by diluting the commercially available 30% solution). After 2 h dilute the reaction mixture with H_2O (using ice cooling if necessary), neutralize, check for completeness of destruction, and discard it.

Destruction of Nitrosamines in Alcohols

Dilute the solution, if necessary, so that the nitrosamine concentration does not exceed 10 mg/mL. Add an equal volume of KOH solution (1 M) and stir the mixture magnetically. For every 100 mL of this solution add 5 g of Ni-Al alloy at such a rate that excessive frothing does not occur. The reaction can be quite exothermic. Do it in a reaction vessel whose volume is at least three times that of the final reaction mixture. Cover the reaction mixture, stir for 24 h, then filter it through a pad of Celite. Neutralize the filtrate, check for completeness of destruction, and discard it. Allow the spent nickel to dry on a metal tray for 24 h (away from flammable solvents) and discard it.

Destruction of Nitrosamines in Dimethyl Sulfoxide

Dilute the mixture with methanol, if necessary, so that the nitrosamine concentration does not exceed 10 mg/mL. Add an equal volume of KOH solution (1 M) and stir the mixture magnetically. For every liter of this solution add 100 g of Ni-Al alloy at such a rate that excessive frothing does not occur. The reaction can be quite exothermic. Do it in a reaction vessel whose volume is at least four times that of the final reaction mixture. Cover the reaction mixture, stir for 24 h, then filter it through a pad of Celite. Neutralize the filtrate, check for completeness of destruction, and discard it. Allow the spent nickel to dry on a metal tray for 24 h (away from flammable solvents) and discard it.

Destruction of Nitrosamines in Olive Oil and Mineral Oil

Dilute olive oil solutions with olive oil and mineral oil solutions with hexane, if necessary, so that the nitrosamine concentration does not exceed 10 mg/mL. Add an equal volume of KOH solution (1 M) and stir the resulting two-phase mixture magnetically. For every 100 mL of this solution add 5 g of Ni-Al alloy at such a rate that excessive frothing does not occur. The reaction can be quite exothermic. Do it in a reaction vessel whose volume is at least four times that of the final reaction mixture. Cover the reaction mixture, stir for 24 h, then filter it through a pad of Celite. Neutralize the filtrate, check for completeness of destruction, and discard it. Allow the spent nickel to dry on a metal tray for 24 h (away from flammable solvents) and discard it.

Destruction of Nitrosamines in Agar Gel

Add the contents of the agar plate ($\sim$ 17 g) to 75 mL of KOH solution (1 M) and stir and warm the mixture until the agar dissolves. Note that many nitrosamines are liable to volatilize under these conditions. When cool add 2 g of Ni-Al alloy slowly enough that excessive frothing does not occur. (If more than 100 mg of nitrosamine is present, increase the weight of alloy and volume of KOH solution proportionately.) Stir the mixture for 24 h, then filter it through a pad of Celite. Neutralize the filtrate, check for completeness of destruction, and discard it. Allow the spent nickel to dry on a metal tray away from flammable solvents for 24 h, then discard it.

Dealing With Spills of Nitrosamines

Remove as much of the spill as possible with an absorbent and cover the remainder with a 47.4 g/L solution of $KMnO_4$ in 3 M H_2SO_4. Leave this mixture overnight, then decolorize with ascorbic acid or sodium ascorbate, neutralize, and remove. If possible, check for completeness of decontamination by taking wipe samples and analyze these samples. Note that this procedure may damage painted surfaces or Formica. Place the absorbent material in a beaker and decontaminate it by one of the methods described above.

Decontamination of Equipment Contaminated with Nitrosamines

A. Rinse five times with an appropriate solvent, then decontaminate this solution using one of the methods described above. To minimize the scale of the final reaction keep the volume of these rinses as small as practicable.

B. Fill the equipment with a 47.4 g/L solution of $KMnO_4$ in 3 M H_2SO_4 (or cover the contaminated surface with this solution). After leaving overnight the solution should still be purple. If not, replace with fresh solution, which should stay purple for at least 1 h. At the end of the reaction decolorize with sodium ascorbate or ascorbic acid, neutralize, check for completeness of destruction, discard the solution, then clean the equipment.

Analytical Procedures

Many procedures have been published for the analysis of nitrosamines. The thermal energy analysis (TEA) detector is specific for nitrosamines and great sensitivity can be achieved using this detector. For many applications the flame ionization detector (FID) provides sufficient sensitivity.

For analysis by gas chromatography[24] a 1.8 m × 2-mm i.d. packed column can be used. For N-nitrosodimethylamine (80°C), N-nitrosodiethylamine (120°C), N-nitrosodiisopropylamine (120°C), N-nitrosodibutylamine (150°C), N-nitrosopyrrolidine (140°C), N-nitrosopiperidine (150°C), and N-nitrosomorpholine (150°C) the packing is 10% Carbowax 20 M + 2% KOH on 80/100 Chromosorb W AW, for N,N'-dinitrosopiperazine (180°C) the packing is 2% Carbowax 20 M + 1% KOH on 80/100 Supelcoport and for N-methyl-N-nitrosoaniline (110°C) the packing is 3% SP 2401-DB on 100/120 Supelcoport. The oven temperatures shown above in parentheses are only a guide and the exact conditions would have to be determined experimentally. High-pressure liquid chromatography conditions for nitrosamines have been described using a 250 × 4.6-mm i.d. column of 10 μ Lichrosorb Si 60 with acetone : isooctane (7:93) flowing at 2 mL/min.[26] A TEA detector was used.

Mutagenicity Assays

The mutagenicity assays were carried out as described on page 4 using tester strains TA98, TA100, TA1530, and TA1535. The reaction mixtures obtained from the Ni-Al alloy degradation of NDMA, NDBA, NPYR, and NPIP were neutralized, diluted with pH 7 buffer in an attempt to avoid toxicity problems (probably from aluminum compounds) and tested. None of the reaction mixtures was mutagenic. Each of the four nitrosamines was mutagenic when tested as the pure compound, whereas the corresponding amines were not mutagenic.[27]

Related Compounds

The procedures listed above should be generally applicable to nitrosamines. Limitations were usually imposed by the matrices in which the nitrosamines were found rather than the nature of the nitrosamines themselves. The only instance where the nature of the nitrosamine made a difference was the use of HBr in glacial acetic acid. In this case, *N*-nitrosopyrrolidine took significantly longer to degrade than the other nitrosamines tested. The conditions given above should, however, degrade *N*-nitrosopyrrolidine.

Nickel-aluminum alloy has also been shown to reduce the related nitramines *N*-nitromorpholine and diisopropylnitramine to the corresponding amines.[28] Destruction of the nitramines was >99.9% and in the final reaction mixtures <0.1% of the theoretical amounts of the related nitrosamines or hydrazines were found.[27] The reaction should be generally applicable to nitramines.

References

1. Other names are dibutylnitrosamine and *N*-butyl-*N*-nitroso-1-butanamine.

2. Other names are methylphenylnitrosamine, *N*-methyl-*N*-nitrosoaniline, and *N*-methyl-*N*-nitrosobenzenamine.

3. International Agency for Research on Cancer. *IARC Monographs on the Evaluation of Carcinogenic Risk of Chemicals to Man.* Volume 1; International Agency for Research on Cancer: Lyon, 1971; pp 95–106.

4. International Agency for Research on Cancer. *IARC Monographs on the Evaluation of the Carcinogenic Risk of Chemicals to Humans.* Volume 17, *Some N-Nitroso Compounds*; International Agency for Research on Cancer: Lyon, 1978; pp 125–175.

5. Reference 3, pp 107–124.

6. Reference 4, pp 83–124.

7. Reference 4, pp 177–189.

8. International Agency for Research on Cancer. *IARC Monographs on the Evaluation of Carcinogenic Risk of Chemicals to Man.* Volume 4, *Some Aromatic Amines, Hydrazine and Related Substances, N-Nitroso Compounds and Miscellaneous Alkylating Agents*; International Agency for Research on Cancer: Lyon, 1974; pp 197–210.

9. Reference 4, pp 51–75.

10. Reference 4, pp 313–326.

11. Reference 4, pp 287–301.

12. Reference 4, pp 263–280.

13. Sax, N.I; Lewis, R.J., Sr. *Dangerous Properties of Industrial Materials*, 7th ed.; Van Nostrand-Reinhold: New York, 1989; p. 2550.

14. Hadidian, Z.; Fredrickson, T.N.; Weisburger, E.K.; Weisburger, J.H.; Glass, R.M.; Mantel, N. Tests for chemical carcinogens. Report on the activity of derivatives of aromatic amines, nitrosamines, quinolines, nitroalkanes, amides, epoxides, aziridines, and purine antimetabolites. *J. Natl. Cancer. Inst.* **1968**, *41*, 985–1036.

15. Michejda, C.J.; Kroeger-Koepke, M.B.; Kovatch, R.M. Carcinogenic effects of sequential administration of two nitrosamines in Fischer 344 rats. *Cancer Res.* **1986**, *46*, 2252–2256.

16. International Agency for Research on Cancer. *IARC Monographs on the Evaluation of the Carcinogenic Risk of Chemicals to Humans.* Volume 27, *Some Aromatic Amines, Anthraquinones and Nitroso Compounds, and Inorganic Fluorides Used in Drinking-water and Dental Preparations;* International Agency for Research on Cancer: Lyon, 1982; pp 213–225.

17. Reference 13, pp 1413–1414.

18. Reference 13, p. 1484.

19. Reference 13, p. 2549.

20. Reference 13, pp 1492–1493.

21. Reference 13, pp 644–645.

22. Reference 13, pp 1458–1459.

23. Reference 13, pp 2353–2354.

24. Lunn, G.; Sansone, E.B.; Keefer,L.K. Safe disposal of carcinogenic nitrosamines. *Carcinogenesis* **1983**, *4*, 315–319.

25. Castegnaro, M.; Eisenbrand, G.; Ellen, G.; Keefer, L.; Klein, D.; Sansone, E. B.; Spincer, D.; Telling, G.; Webb, K., Eds. *Laboratory Decontamination and Destruction of Carcinogens in Laboratory Wastes: Some* N-*Nitrosamines*; International Agency for Research on Cancer: Lyon, 1982 (IARC Scientific Publications No. 43).

26. Goff, U. High-performance liquid chromatography of volatile nitrosamines. In *Environmental Carcinogens Selected Methods of Analysis.* Volume 6, *N-Nitroso Compounds*; Egan, H., Preussmann, R., Eisenbrand, G., Spiegelhalder, B., O'Neill, I.K., Bartsch, H., Eds.; International Agency for Research on Cancer: Lyon, 1983 (IARC Scientific Publications No. 45); pp 389–394.

27. Lunn, G. Unpublished results.

28. Lunn, G.; Sansone, E.B.; Keefer, L.K. General cleavage of N-N and N-O bonds using nickel/aluminum alloy. *Synthesis* **1985**, 1104–1108.

NITROSOUREA DRUGS

> **CAUTION!** Refer to safety considerations section on page 6 before starting any of these procedures.

The nitrosourea drugs considered in this section are all antineoplastic agents and they all have an N-NO functionality in common. They are all crystalline solids and are moderately soluble in alcohols. Their H_2O-solubility varies.

The compounds considered are:

BCNU (**I**) N,N'-Bis(2-chloroethyl)-N-nitrosourea;[1] mp 30–32°C; solubility in H_2O 4 mg/mL

CCNU (**II**) N-(2-Chloroethyl)-N'-cyclohexyl-N-nitrosourea;[2] mp 90°C; solubility in H_2O <0.05 mg/mL

(**I**)   (**II**)

199

STZ (**III**, R = CH$_3$)   Streptozotocin;[3]
                                   mp 115°C; soluble in H$_2$O and ketones, not sol-
                                   uble in other organic solvents

CTZ (**III**, R = CH$_2$CH$_2$Cl)   Chlorozotocin;[4]
                                   mp 140–141°C; soluble in H$_2$O

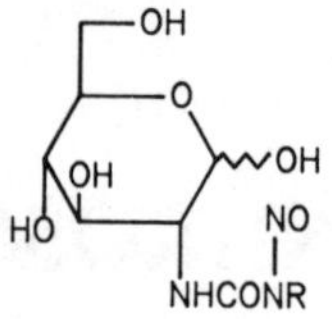

(III)

PCNU (**IV**)   N-(2-Chloroethyl)-N′-(2,6-dioxo-3-piperidinyl)-
                                   N-nitrosourea;
                                   mp 130–131°C; solubility in H$_2$O <1 mg/mL,
                                   soluble in acetone

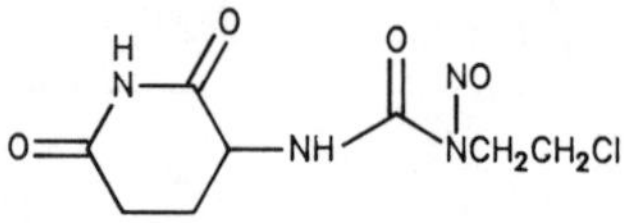

(IV)

Methyl CCNU (**V**)   N-(2-Chloroethyl)-N′-(4-methylcyclohexyl)-N-
                                   nitrosourea;[5]
                                   mp 68–69°C; solubility in H$_2$O 0.09 mg/mL,
                                   soluble in dimethyl sulfoxide

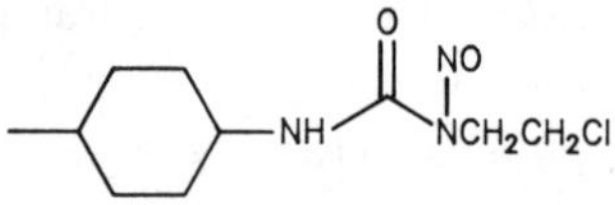

(V)

The compounds BCNU,[6–8] CCNU,[8–10] methyl CCNU,[8,11] CTZ,[12] and STZ[13] are carcinogenic in experimental animals and the International Agency for Research on Cancer has stated that BCNU,[6,8] CCNU,[9] methyl CCNU,[8] and STZ[13] should be regarded as presenting a carcinogenic risk to humans. In addition, BCNU[14] and CCNU[15] are teratogens and have toxic effects on the blood and so on, STZ is a teratogen and affects the liver and kidneys,[16] and methyl CCNU can affect the kidneys.[17] All of these compounds are mutagenic.[18]

Principles of Destruction

Streptozotocin is degraded with saturated sodium bicarbonate ($NaHCO_3$) solution (<0.012% remains) to give methanol.[18] The other drugs are degraded by reduction with nickel-aluminum alloy (Ni-Al) in potassium hydroxide (KOH) solution (<0.2% remains).[18] To obtain nonmutagenic products from the reduction of BCNU it is necessary to increase the ratio of reductant to substrate.[18] The products obtained were ethanol and cyclohexylamine (from CCNU) and 4-methylcyclohexylamine (from methyl CCNU). No trace of diazoethane or 2-chlorodiazoethane were found from CCNU and no trace of diazomethane was found from STZ. The use of hydrogen bromide in glacial acetic acid has been proposed for the destruction of these compounds,[19] but in a number of cases the product was found to be the denitrosated compound,[18] which was also a mutagen so this method cannot be recommended as a destruction procedure.

Destruction Procedures

Destruction of Bulk Quantities of CCNU, CTZ, PCNU, and Methyl CCNU

Dissolve the nitrosourea in methanol so that the concentration does not exceed 10 mg/mL, then add an equal volume of 2 M KOH solution. For every 20 mL of this basified solution add 1 g of Ni-Al alloy. Add quantities of >5 g in portions to prevent the reaction from frothing too violently. Perform the reaction in a vessel at least three times larger than the final volume as some foaming may occur. Stir the mixture overnight, then filter through a pad of Celite. Neutralize the filtrate, check for completeness of destruction, and discard it. Allow the spent nickel, which is filtered off, to dry on a metal tray away from flammable solvents for 24 h, then discard it with the solid waste.

Destruction of Bulk Quantities of BCNU

Dissolve 100 mg of BCNU in 3 mL of ethanol and add 27 mL of H_2O. To this solution add 30 mL of 2 M KOH solution and 3 g of Ni-Al alloy. Add quantities of Ni-Al alloy in excess of 5 g in portions to prevent the reaction from frothing too violently. Perform the reaction in a vessel at least three times larger than the final volume as some foaming may occur. Stir the mixture overnight, then filter through a pad of Celite. Neutralize the filtrate, check for completeness of destruction, and discard it. Allow the spent nickel, which is filtered off, to dry on a metal tray away from flammable solvents for 24 h, then discard it with the solid waste.

Destruction of Pharmaceutical Preparations of BCNU, CCNU, and CTZ

BCNU	The pharmaceutical preparation consists of 100 mg of drug in 3 mL of ethanol to which is added 27 mL of H_2O.
CCNU	Open the capsules and allow the shells to remain in the reaction vessel. For every capsule (100 mg) add 10 mL of methanol.
CTZ	The pharmaceutical preparation consists of 50 mg of drug in 5 mL of saline solution.

To each of these solutions add an equal volume of 2 M KOH solution. For every 20 mL of this basified solution add 1 g of Ni-Al alloy. Add quantities of more than 5 g in portions to prevent the reaction from frothing too violently. Perform the reaction in a vessel at least three times larger than the final volume as some foaming may occur. Stir the mixture overnight, then filter through a pad of Celite. Neutralize the filtrate, check for completeness of destruction, and discard it. Allow the spent nickel, which is filtered off, to dry on a metal tray away from flammable solvents for 24 h, then discard it with the solid waste.

Destruction of STZ

Take up bulk quantities in H_2O so that the concentration does not exceed 100 mg/mL. If necessary, dilute pharmaceutical preparations with H_2O so that they do not exceed 100 mg/mL. For each volume of STZ solution add five times the volume of saturated $NaHCO_3$ solution, allow the mixture to stand overnight, check for completeness of destruction, and discard it. Prepare saturated $NaHCO_3$ solution by mixing $NaHCO_3$ and H_2O in a container. Shake the container occasionally. If solid persists, the solution is saturated; if not, add more $NaHCO_3$.

Analytical Procedures

Analysis was by HPLC using a 250 × 4.6-mm i.d. column of Microsorb C8. The injection volume was 20 μL and the mobile phase flowed at 1 mL/min. Ultraviolet detection at 254 nm was used. For BCNU, CCNU, PCNU, and methyl CCNU methanol and a 3.5 mM $(NH_4)H_2PO_4$ buffer were used; for BCNU and PCNU the MeOH : buffer ratio was 50:50 and for CCNU and methyl CCNU it was 75:25. For CTZ methanol : 20 mM KH_2PO_4 buffer (4:96) was used and for STZ pure 20 mM KH_2PO_4 buffer was used. On our equipment these mobile phase combinations were found to give reasonable retention times (4–14 min). It was frequently advantageous to add some KH_2PO_4 buffer to an aliquot of the neutralized reaction mixture and to centrifuge before analysis. This removed salts that could clog the chromatograph.

Gas chromatography (GC) using a 1.8 m × 2-mm i.d. glass column packed with 10% Carbowax 20 M + 2% KOH on 80/100 Chromosorb W AW was used to determine the products of these reactions. The injection temperature was 200°C, except where shown, and the flame ionization detector operated at 300°C. The oven temperature was 60°C and approximate retention times were methanol (1.7 min), ethanol (2.2 min), cyclohexylamine (13 min), and 4-methylcyclohexylamine (17 min).

To check for diazoalkanes the reaction mixture was swept with nitrogen into ether containing either valeric acid or acetic acid. The ether solution was washed with $NaHCO_3$ solution, then analyzed by GC with the equipment described above for the presence of esters generated by the reaction of diazoalkanes with the carboxylic acids. Thus, CCNU might give 2-chlorodiazoethane which would produce 2-chloroethyl acetate (retention time 5 min, oven temperature 110°C, injection temperature 110°C) or diazoethane (produced by Ni-Al dechlorination of the starting material or 2-chlorodiazoethane), which would produce ethyl acetate (retention time 2.2 min, oven temperature 60°C) and STZ might give diazomethane which would produce methyl valerate (retention time 4 min, oven temperature 70°C). No diazoalkanes were detected.

Mutagenicity Assays[18]

The mutagenicity assays were carried out as described on page 4 using tester strains TA98, TA100, TA1530, and TA1535. To avoid cell toxicity problems it was generally necessary to mix an aliquot of the neutralized reaction mixture with an equal volume of pH 7 buffer before testing. The final reaction mixtures [tested at a level corresponding to 0.25 mg (0.085 mg for BCNU) undegraded material per plate] were not mutagenic. The reaction mixtures from the STZ degradations were tested without using buffer at a level corresponding to 1.7 mg of undegraded product and were not mutagenic. All of the drugs were tested in dimethyl sulfoxide solution and were found to be mutagenic. None of the products detected were found to be mutagenic.

Related Compounds

The Ni-Al alloy technique described above should be applicable to compounds of the general form R-NH-CO-N(NO)-CH$_2$CH$_2$Cl, but the procedure should be thoroughly validated. The problems encountered with BCNU indicate that sometimes a higher reductant : substrate ratio may be required.

References

1. Other names are Carmustine, BiCNU, and Nitrumon.

2. Other names are Lomustine, 1-(2-chloroethyl)-3-cyclohexyl-1-nitrosourea, Belustine, Cecenu, and CeeNU.

3. Other names are 2-deoxy-2-([(methylnitrosoamino)carbonyl]amino)-D-glucopyranose, streptozocin, 2-deoxy-2-(3-methyl-3-nitrosoureido)-D-glucopyranose, N-D-glucosyl-(2)-N'-nitrosomethylurea, and Zanosar.

4. Other names are 2-[([(2-chloroethyl)nitrosoamino]carbonyl)amino]-2-deoxy-D-glucose, 2-[3-(2-chloroethyl)-3-nitrosoureido]-2-deoxy-D-glucosopyranose, 1-(2-chloroethyl)-1-nitroso-3-(D-glucos-2-yl)urea, and DCNU.

5. Another name is Semustine.

6. International Agency for Research on Cancer. *IARC Monographs on the Evaluation of the Carcinogenic Risk of Chemicals to Humans.* Volume 26, *Some Antineoplastic and Immunosuppressive Agents*; International Agency for Research on Cancer: Lyon, 1981; pp 79-95.

7. International Agency for Research on Cancer. *IARC Monographs on the Evaluation of the Carcinogenic Risk of Chemicals to Humans, Supplement No. 4, Chemicals, Industrial Processes and Industries Associated with Cancer in Humans. IARC Monographs, Volumes 1 to 29*; International Agency for Research on Cancer: Lyon, 1982; pp 63–64.

8. International Agency for Research on Cancer. *IARC Monographs on the Evaluation of the Carcinogenic Risk of Chemicals to Humans, Supplement No. 7, Overall Evaluations of Carcinogenicity: An Updating of* IARC Monographs *Volumes 1 to 42*; International Agency for Research on Cancer: Lyon, 1987; pp 150–152.

9. Reference 6, pp 137–149.

10. Reference 7, pp 83–84.

11. Weisburger, E.K. Bioassay program for carcinogenic hazards of cancer chemotherapeutic agents. *Cancer* **1977**, *40*, 1935–1949.

12. Habs, M.; Eisenbrand, G.; Schmähl, D. Carcinogenic activity in Sprague-Dawley rats of 2-[3-(2-chloroethyl)-3-nitrosoureido]-D-glucopyranose (chlorozotocin). *Cancer Lett.* **1979**, *8*, 133–137.

13. International Agency for Research on Cancer. *IARC Monographs on the Evaluation of the Carcinogenic Risk of Chemicals to Humans.* Volume 17, *Some N-Nitroso Compounds*; International Agency for Research on Cancer: Lyon, 1978; pp 337–349.

14. Sax, N.I; Lewis, R.J., Sr. *Dangerous Properties of Industrial Materials*, 7th ed.; Van Nostrand-Reinhold: New York, 1989; pp 476–477.

15. Reference 14, p. 817.

16. Reference 14, p. 3121.

17. Reference 14, pp 822–823.

18. Lunn, G.; Sansone, E.B.; Andrews, A.W.; Hellwig, L.C. Degradation and disposal of some antineoplastic drugs. *J. Pharm. Sci.* **1989**, *78*, 652–659.

19. Castegnaro, M.; Adams, J.; Armour, M-. A.; Barek, J.; Benvenuto, J.; Confalonieri, C.; Goff, U.; Ludeman, S.; Reed, D.; Sansone, E. B.; Telling, G., Eds. *Laboratory Decontamination and Destruction of Carcinogens in Laboratory Wastes: Some Antineoplastic Agents*; International Agency for Research on Cancer: Lyon, 1985 (IARC Scientific Publications No. 73).

ORGANIC NITRILES

> **CAUTION!** Refer to safety considerations section on page 6 before starting any of these procedures.

Little work has been done on the chemical degradation of these compounds but some results have been obtained for simple compounds. The following compounds were degraded with nickel-aluminum (Ni-Al) alloy in dilute base:[1] acetonitrile,[2] 3-ethoxypropionitrile, benzonitrile,[3] and benzyl cyanide.[4] Acetonitrile (bp 82°C), 3-ethoxypropionitrile (bp 172°C), benzonitrile (bp 188°C), and benzyl cyanide (bp 233–234°C) are volatile liquids. Acetonitrile is moderately toxic, produces convulsions, nausea, and vomiting, and is a teratogen.[5] Benzyl cyanide has been known to explode with sodium hypochlorite.[6]

Principle of Destruction

These compounds were reduced with Ni-Al alloy in dilute base to give the corresponding amine in 67–86% yield. Destruction was >99% in all cases.

Destruction Procedure

Take up 0.5 g of the nitrile in 50 mL of H_2O, then add 50 mL of 1 M potassium hydroxide (KOH) solution. Stir this mixture and add 5 g of Ni-Al alloy in portions to avoid frothing. Stir the reaction mixture overnight, then filter through a pad of Celite. Neutralize the filtrate, check for completeness of destruction, and discard it. Place the spent nickel on a metal tray, allow it to dry away from flammable solvents for 24 h, and discard it.

Analytical Procedures

For analysis by gas chromatography[1] a 1.8 m × 2-mm i.d. packed column can be used together with flame ionization detection. The injection temperature was 200°C, the detector temperature was 300°C, and the carrier gas was nitrogen flowing at 30 mL/min. For acetonitrile (100°C) and 3-ethoxypropionitrile (100°C) the packing was 10% Carbowax 20 M + 2% KOH on 80/100 Chromosorb W AW and for benzonitrile (120°C) and benzyl cyanide (120°C) the packing was 2% Carbowax 20 M + 1% KOH on 80/100 Supelcoport. The oven temperatures shown above in parentheses are only a guide and the exact conditions would have to be determined experimentally.

Related Compounds

Reduction with Ni-Al alloy does not degrade inorganic cyanides.[1] These should be degraded as described in the cyanides and cyanogen bromide section, page 77. A number of aryl nitriles have been reduced to the corresponding amines with Ni-Al alloy,[7,8] but the complete destruction of the starting material has not been established. Full validation should be carried out before this process is used on a routine basis.

References

1. Lunn, G. Unpublished observations.
2. Other names are ethyl nitrile, methanecarbonitrile, cyanomethane, methyl cyanide, and ethanenitrile.
3. Other names are benzoic acid nitrile, cyanobenzene, and phenyl cyanide.
4. Other names are benzeneacetonitrile, benzyl cyanide, benzyl nitrile, α-cyanotoluene, ω-cyanotoluene, phenylacetonitrile, and α-tolunitrile.

5. Sax, N.I; Lewis, R.J., Sr. *Dangerous Properties of Industrial Materials*, 7th ed.; Van Nostrand-Reinhold: New York, 1989; p. 27.

6. Bretherick, L., Ed. *Hazards in the Chemical Laboratory*, 4th ed.; Royal Society of Chemistry: London, 1986; pp 449–450.

7. Staskun, B.; van Es, T. Reductions with Raney alloy in alkaline solution. *J. Chem. Soc. (C)* **1966**, 531–532.

8. Kametani, T.; Nomura, Y. Studies on a catalyst. II. Reduction of nitrogen compounds by Raney nickel alloy and alkali solution. 2. Synthesis of amines by reduction of nitriles. *J. Pharm. Soc. Japan* **1954**, *74*, 889–891; *Chem. Abstr.* **1956**, *50*, 2467f.

OSMIUM TETROXIDE

> **CAUTION!** Refer to safety considerations section on page 6 before starting any of these procedures.

Osmium tetroxide (OsO_4, osmic acid) is a low-melting solid, mp 39.5–41°C, bp 130°C. It is quite volatile and its vapors irritate and burn the eyes severely[1] and affect the lungs.[2] It is widely used in synthetic organic chemistry and in electron microscopy laboratories.

Principle of Disposal

Osmium tetroxide reacts with double bonds to form a very stable diester. In this form the OsO_4 is no longer volatile. Corn oil contains a large proportion of double bonds and it is an effective agent for the neutralization of OsO_4.[3] Commercially available Mazola corn oil was used in the tests. Although this procedure eliminates hazards due to the volatility of OsO_4, the material still contains osmium and it should be disposed of as waste containing heavy metals. It has been reported that OsO_4, can be reduced to the dioxide by reacting it with an olefin, bubbling hydrogen sulfide through the solution and removing the osmium dioxide by filtration.[4]

211

Disposal Procedures

Bulk Quantities and Residues in Containers

Place corn oil in the container. Test for completeness of reaction.

Aqueous Solutions (2%)

Allow to react with twice the volume of corn oil. Test for completeness of reaction.

Spills

Two hundred g of absorbent granules (for example, cat litter) absorbs 100 mL of corn oil. Use this mixture to neutralize a spill of 50 mL of 2% OsO_4. Test for completeness of reaction then remove the granules. The corn oil-absorbent granules mixture can be kept in a tightly sealed plastic bag for at least 1 month with no loss of effectiveness. Laboratories in which OsO_4 is in routine use should keep at least one bag on hand.

Analytical Procedures

Either a glass cover slip coated in corn oil or a piece of filter paper soaked in corn oil was suspended over the solution. Blackening indicated that OsO_4 was still present.

Related Compounds

This method is specific for OsO_4.

References

1. Bretherick, L., Ed. *Hazards in the Chemical Laboratory*, 4th ed.; Royal Society of Chemistry: London, 1986; p. 434.
2. Sax, N.I; Lewis, R.J., Sr. *Dangerous Properties of Industrial Materials*, 7th ed.; Van Nostrand-Reinhold: New York, 1989; pp 2622–2623.
3. Cooper, K. Neutralization of osmium tetroxide in case of accidental spillage and for disposal. *Bulletin of the Microscopical Society of Canada* **1980**, *8*, 24–28.
4. Armour, M-.A.; Browne, L.M.; Weir, G.L., Eds. *Hazardous Chemicals. Information and Disposal Guide,* 3rd ed.; University of Alberta: Edmonton, Alberta, 1987; p. 263.

PEROXIDES

CAUTION! Refer to safety considerations section on page 6 before starting any of these procedures.

CAUTION! Peroxides are frequently formed in certain organic solvents such as ethers. These compounds are dangerously unstable and actions such as removing the container cap may cause them to explode. The help of people specially trained to deal with explosives should be sought in these cases.

Some relatively stable peroxides are used in organic chemistry, particularly for the initiation of polymerization reactions. If care is exercised, the use of these compounds should not be particularly hazardous although they may explode when subjected to shock or exposed to heat. They are powerful oxidizers and may react violently with reducing agents.[1]

Destruction Procedures[2]

Diacyl Peroxides

Dissolve 3.3 g of sodium iodide (NaI) or 3.65 g of potassium iodide (KI) in 70 mL of glacial acetic acid. Stir this mixture at room temperature and

213

slowly add 0.01 mol of the peroxide. The mixture darkens because iodine is formed. After 30 min discard the mixture.

Dialkyl peroxides

Dissolve 3.3 g of NaI or 3.65 g of KI in 70 mL of glacial acetic acid and add 1 mL of concentrated hydrochloric acid as an accelerator. Stir this mixture at room temperature and slowly add 0.01 mol of the peroxide. The mixture darkens because iodine is formed. Heat the mixture slowly (over 30 min) to 90–100°C on a steam bath and hold it at that temperature for 5 h, then discard it.

References

1. Sax, N.I; Lewis, R.J., Sr. *Dangerous Properties of Industrial Materials*, 7th ed.; Van Nostrand-Reinhold: New York, 1989; p. 2693.
2. National Research Council, Committee on Hazardous Substances in the Laboratory. *Prudent Practices for Disposal of Chemicals from Laboratories;* National Academy Press: Washington, DC, 1983; p. 76.

PHOSPHORUS AND
PHOSPHORUS PENTOXIDE

CAUTION! Refer to safety considerations section on page 6 before starting any of these procedures.

Phosphorus is used in the chemical laboratory. White phosphorus (sometimes called yellow phosphorus) is pyrophoric and can explode when it reacts with a variety of chemicals.[1] It is a poison and has a number of severe health effects including necrosis of the jaw.[1] Red phosphorus is not pyrophoric, but it is flammable and can explode when mixed with a variety of compounds.[2] Phosphorus pentoxide (phosphoric anhydride, diphosphorus pentoxide, or P_2O_5) is corrosive and reacts violently with H_2O.[3] It is used as a drying or dehydrating agent in the laboratory.

Principles of Destruction

White phosphorus is oxidized by copper(II) sulfate to phosphoric acid and red phosphorus is oxidized to phosphoric acid by potassium chlorate. Phosphorus pentoxide is hydrolyzed to phosphoric acid.

215

Destruction Procedures[4]

White Phosphorus

Cut 5 g of white phosphorus under H_2O into pellets that are no more than 5 mm across and add these pellets to 800 mL of 1 M cupric sulfate solution. Allow the reaction mixture to stand in a 2-L beaker in a hood for about a week. Stir occasionally. If one of the larger black pellets is cut under H_2O and no waxy white phosphorus is observed, the reaction is complete. Filter off the precipitate and, while keeping it wet, add to 500 mL of 5.25% sodium hypochlorite solution. Stir this mixture for 1 h to oxidize any copper phosphide to copper phosphate. Dispose of the final reaction mixture in an appropriate fashion. Use fresh sodium hypochlorite solution (see assay procedure below).

Red Phosphorus

Add 5 g of red phosphorus to a solution of 33 g of potassium chlorate in 2 L of 0.5 M sulfuric acid and heat the mixture under reflux until all the phosphorus has dissolved ($\sim$ 5–10 h). After cooling to room temperature add 14 g of sodium bisulfite to reduce the excess chlorate and discard the reaction mixture.

Phosphorus Pentoxide

Gradually add P_2O_5 to a stirred mixture of H_2O and crushed ice. Before discarding it ensure that no chunks of unreacted P_2O_5 are left.

Assay of Sodium Hypochlorite Solution

Sodium hypochlorite solutions tend to deteriorate with time so they should be periodically checked for the amount of active chlorine they contain. Pipette 10 mL of sodium hypochlorite solution into a 100-mL volumetric flask and fill to the mark with distilled H_2O. Pipette 10 mL of this solution into a conical flask containing 50 mL of distilled H_2O, 1 g of potassium iodide, and 12.5 mL of 2 M acetic acid. Titrate this solution against 0.1 N sodium thiosulfate solution using starch as an indicator. Each 1 mL of the sodium thiosulfate solution corresponds to 3.545 mg of active chlorine. The sodium hypochlorite solution used in these degradation reactions should contain 25–30 g of active chlorine/L.

References

1. Sax, N.I; Lewis, R.J., Sr. *Dangerous Properties of Industrial Materials*, 7th ed.; Van Nostrand-Reinhold: New York, 1989; pp 2776–2777.

2. Reference 1, pp 2775–2776.

3. Reference 1, p. 2779.

4. National Research Council, Committee on Hazardous Substances in the Laboratory. *Prudent Practices for Disposal of Chemicals from Laboratories;* National Academy Press: Washington, DC, 1983; pp 92–93.

PICRIC ACID

CAUTION! Refer to safety considerations section on page 6 before starting any of these procedures.

Picric acid (2,4,6-trinitrophenol, carbazotic acid, 2-hydroxy-1,3,5-trinitro-benzene, nitroxanthic acid, phenol trinitrate, or picronitric acid) is used in the chemical laboratory for preparing picrates for the characterization of organic compounds. It is also used in histological stains. When wet it is quite stable but in the dry form it is an explosive. It can form explosive salts with many metals. It can cause local and systemic reactions with a variety of symptoms.[1] Only wet picric acid should be degraded. Seek professional help for dry picric acid.

Principles of Destruction

The nitro groups of picric acid may be reduced using sodium sulfide or tin in hydrochloric acid (HCl). Although the product should in theory be 2,4,6-triaminophenol it is likely that reduction will not be complete. In addition, after reduction has ceased air oxidation will convert the product

219

to a mixture of nitroamines, various dimers, and other hazardous compounds. Thus reduction of picric acid will probably convert it to a mixture of hazardous compounds but at least they will not be explosive. The products of these reactions should be disposed of as hazardous waste.

Destruction Procedures

Destruction of Bulk Quantities

A. Dissolve 0.13 g of sodium hydroxide (NaOH) in 25 mL of H_2O, then add 2.7 g of sodium sulfide. When the sodium sulfide has dissolved, add 1 g of picric acid.[2] When the reaction appears to be complete dispose of the mixture with the hazardous waste.

  B. Stir 1 g of picric acid, 10 mL of H_2O, and 4 g of granular tin in a flask that is cooled in an ice bath.[3] Add concentrated HCl (15 mL), cautiously at first, and, when addition is complete, allow the reaction mixture to warm to room temperature. Reflux the reaction for 1 h, cool, and filter it. Wash the unreacted tin with 10 mL of 2 M HCl and neutralize the filtrate with 10% NaOH. Refilter to remove tin chloride and discard the filtrate as hazardous aqueous waste. Discard the unreacted tin and the tin chloride.

Decontamination of Dilute Aqueous Solutions[3]

Dilute the aqueous solution of picric acid with H_2O, if necessary, so that the concentration does not exceed 0.4%, then for each 100 mL of solution add 2 mL of concentrated HCl to bring the pH to 2. Add granular tin (30 mesh, 1 g) and allow the mixture to stand at room temperature. After ~ 14 days the picric acid is completely degraded. Dispose of the reaction mixture as hazardous aqueous waste.

Analytical Procedures[3]

Picric acid can be determined by thin-layer chromatography on silica gel. The eluant is methanol:toluene:glacial acetic acid (8:45:4) and the picric acid forms a bright yellow spot of $R_f \sim 0.3$. Iodine vapor will increase the sensitivity of the procedure.

References

1. Sax, N.I; Lewis, R.J., Sr. *Dangerous Properties of Industrial Materials*, 7th ed.; Van Nostrand-Reinhold: New York, 1989; p. 2789.

2. Manufacturing Chemists Association. *Laboratory Waste Disposal Manual*; Manufacturing Chemists Association: Washington DC, 1973; p. 133.

3. Armour, M-.A.; Browne, L.M.; Weir, G.L., Eds. *Hazardous Chemicals. Information and Disposal Guide,* 3rd ed.; University of Alberta: Edmonton, Alberta, 1987; p. 317.

POLYCYCLIC AROMATIC
HYDROCARBONS

> **CAUTION!** Refer to safety considerations section on
> page 6 before starting any of these procedures.

In a recent international collaborative study the destruction of the following polycyclic aromatic hydrocarbons (PAH) was investigated:[1]

Benz[a]anthracene[2]	BA	**(I)**
Benzo[a]pyrene[3]	BP	**(II)**
7-Bromomethylbenz[a]anthracene	BrMBA	**(III)**
Dibenz[a,h]anthracene[4]	DBA	**(IV)**
7,12-Dimethylbenz[a]anthracene[5]	DMBA	**(V)**
3-Methylcholanthrene[6]	3-MC	**(VI)**

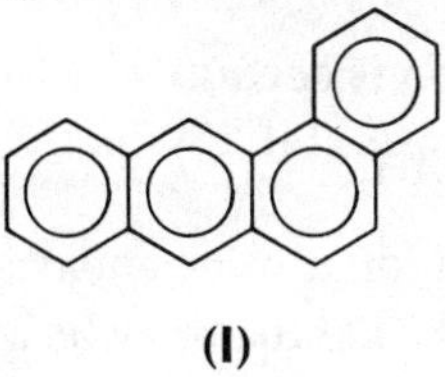

(I)

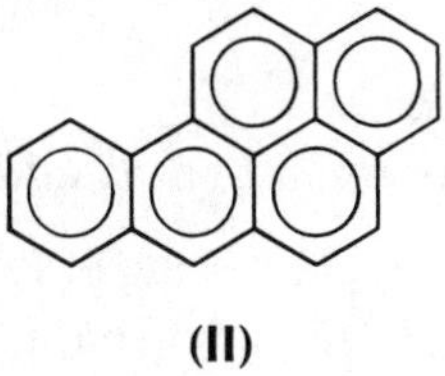

(II)

223

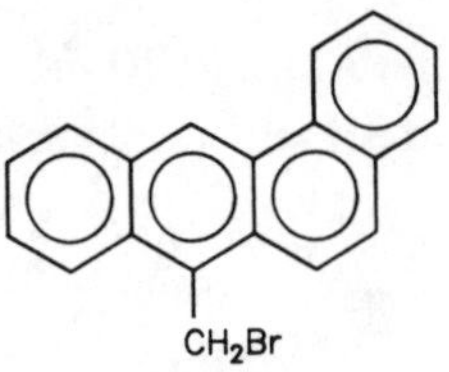

CH$_2$Br

(III)

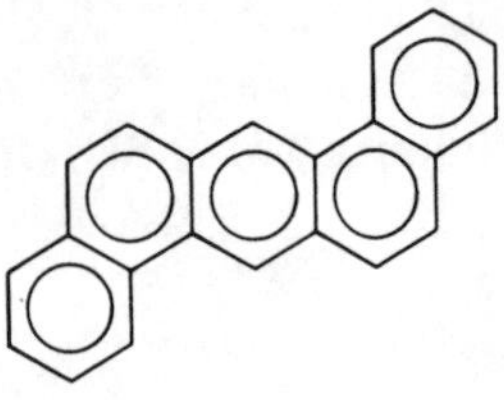

(IV)

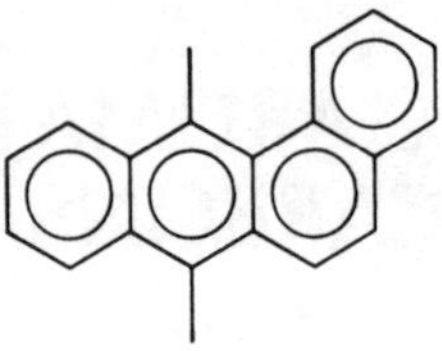

(V)

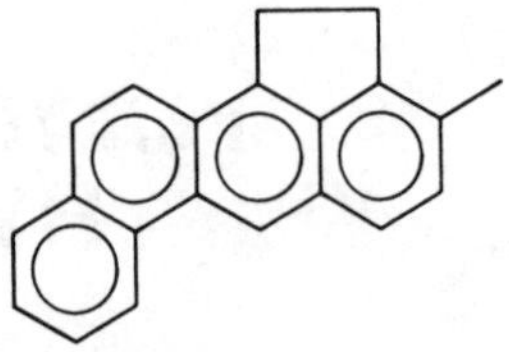

(VI)

These compounds are all high-melting solids (mp >120°C), which are soluble in most organic solvents [for example, benzene, toluene, cyclohexane, acetone, dimethylformamide (DMF), dimethyl sulfoxide (DMSO)], slightly soluble in alcohols, but only soluble in H$_2$O at the microgram per liter level. The compounds BA,[7,8] BP,[9,10] BrMBA,[11] DBA,[12,13] DMBA,[14,15] and 3-MC[16–18] cause cancer in experimental animals. While there is no direct evidence that they cause cancer in humans, coal tar and other materials known to be carcinogenic to humans may contain these PAH so they should all be regarded as potential human carcinogens. The compounds BP,[19] DMBA,[20] and 3-MC[21] are teratogens. Polycyclic aromatic hydrocarbons are widely used in cancer research laboratories and they are also found in the environment as products of combustion.

Principles of Destruction

Polycyclic aromatic hydrocarbons may be destroyed by oxidation with potassium permanganate in sulfuric acid (KMnO$_4$ in H$_2$SO$_4$) or by dissolution in concentrated H$_2$SO$_4$.[1] The products of these reactions have not been determined. Destruction efficiency was >99% in each case.

Destruction Procedures

Destruction of Bulk Quantities of PAH

A. For every 5 mg of PAH add 2 mL of acetone and ensure that the PAH is completely dissolved, including any PAH that may be adhering to the walls of the container. For every 5 mg of PAH add 10 mL of a 0.3 M KMnO$_4$ solution in 3 M H$_2$SO$_4$ (freshly prepared) and swirl the mixture and allow to react for 1 h. The purple color should be maintained during this reaction time. If it is not, add more KMnO$_4$ solution until the reaction mixture remains purple for 1 h. At the end of the reaction decolorize the solution with ascorbic acid, neutralize, test for completeness of destruction, and discard it.

Note. Use *at least* 1 mL of acetone and *at least* 10 mL of KMnO$_4$.

B. For every 5 mg of PAH add 2 mL of DMSO and ensure that the PAH is completely dissolved, including any PAH that may be adhering to the walls of the container. For every 5 mg of PAH add 10 mL of concentrated H$_2$SO$_4$ (**Caution!** Exothermic process) and swirl the mixture and allow to react for at least 2 h. At the end of the reaction cautiously add the solution to at least three times the volume of cold H$_2$O (using an ice bath if desired as this is a very exothermic process), neutralize, test for completeness of destruction, and discard it.

Destruction of PAH in Organic Solvents (Except DMSO and DMF)

A. Remove the solvent by evaporation under reduced pressure using a rotary evaporator. For every 5 mg of PAH add 2 mL of acetone and ensure that the PAH is completely dissolved, including any PAH that may be adhering to the walls of the container. For every 5 mg of PAH add 10 mL of a 0.3 M KMnO$_4$ solution in 3 M H$_2$SO$_4$ (freshly prepared) and swirl the mixture and allow to react for 1 h. The purple color should be maintained during this reaction time. If it is not, add more KMnO$_4$ solution until the reaction mixture remains purple for 1 h. At the end of the reaction decolorize the solution with ascorbic acid, neutralize, test for completeness of destruction, and discard it.

Note. Use *at least* 1 mL of acetone and *at least* 10 mL of KMnO$_4$.

B. Remove the solvent by evaporation under reduced pressure using a rotary evaporator. For every 5 mg of PAH add 2 mL of DMSO and ensure that the PAH is completely dissolved, including any PAH that may be adhering to the walls of the container. For every 5 mg of PAH add 10 mL of concentrated H$_2$SO$_4$ (**Caution!** Exothermic process) and swirl the mixture and allow to react for at least 2 h. At the end of the reaction cautiously add the solution to at least three times the volume of cold H$_2$O (using an ice bath if desired as this is a very exothermic process), neutralize, test for completeness of destruction, and discard it.

Destruction of PAH in DMF

For each 10 mL of DMF solution add 10 mL of H_2O and 20 mL of cyclohexane, shake and allow to separate. Extract the lower, aqueous layer twice more with 20 mL portions of cyclohexane, then combine the cyclohexane layers and remove the solvent by evaporation under reduced pressure using a rotary evaporator. For every 5 mg of PAH add 2 mL of acetone and ensure that the PAH is completely dissolved, including any PAH that may be adhering to the walls of the container. For every 5 mg of PAH add 10 mL of a 0.3 M $KMnO_4$ solution in 3 M H_2SO_4 (freshly prepared) and swirl the mixture and allow to react for 1 h. The purple color should be maintained during this reaction time. If it is not, add more $KMnO_4$ solution until the reaction mixture remains purple for 1 h. At the end of the reaction decolorize the solution with ascorbic acid, neutralize, test for completeness of destruction, and discard it.

Note. Use *at least* 1 mL of acetone and *at least* 10 mL of $KMnO_4$.

Destruction of PAH in DMSO

A. For each 10 mL of DMSO solution add 5 mL of H_2O and 20 mL of cyclohexane, shake, and allow to separate. Extract the lower, aqueous layer twice more with 20 mL portions of cyclohexane, then combine the cyclohexane layers and remove the solvent by evaporation under reduced pressure using a rotary evaporator. For every 5 mg of PAH add 2 mL of acetone and ensure that the PAH is completely dissolved, including any PAH that may be adhering to the walls of the container. For every 5 mg of PAH add 10 mL of a 0.3 M $KMnO_4$ solution in 3 M H_2SO_4 (freshly prepared) and swirl the mixture and allow to react for 1 h. The purple color should be maintained during this reaction time. If it is not, add more $KMnO_4$ solution until the reaction mixture remains purple for 1 h. At the end of the reaction decolorize the solution with ascorbic acid, neutralize it, test it for completeness of destruction, and discard it.

Note. Use *at least* 1 mL of acetone and *at least* 10 mL of $KMnO_4$.

B. Dilute the solution with more DMSO, if necessary, so that the PAH concentration does not exceed 2.5 mg/mL. For every 2 mL of DMSO add 10 mL of concentrated H_2SO_4 (**Caution!** Exothermic process) and swirl the mixture and allow to react for at least 2 h. At the end of the reaction cautiously add the solution to at least three times the volume of cold H_2O (using an ice bath if desired as this is a very exothermic process), neutralize, test for completeness of destruction, and discard it.

Destruction of PAH in Water

Because these compounds are so insoluble, only trace amounts are likely to be present. Add enough $KMnO_4$ to make a 0.3 M solution and enough H_2SO_4 to make a 3 M solution and swirl the mixture and allow it to react for 1 h. The purple color should be maintained during this reaction time. If it is not, add more $KMnO_4$ solution until the reaction mixture remains purple for 1 h. At the end of the reaction decolorize the solution with ascorbic acid, neutralize, test for completeness of destruction, and discard it.

Destruction of PAH in Oil

For each 5 mL of oil solution add 20 mL of 2-methylbutane and 20 mL of acetonitrile, shake for at least 1 minute (**Caution!** Pressure may develop) and allow the layers to separate. Extract the upper, hydrocarbon layer four more times with 20 mL portions of acetonitrile. If necessary, add a 10 mL portion of 2-methylbutane after the second extraction to avoid inversion of the layers caused by evaporation of the 2-methylbutane. Combine the acetonitrile layers and wash them with 20 mL of 2-methylbutane (discard the wash), then remove the solvent by evaporation under reduced pressure using a rotary evaporator (H_2O bath temperature 25–30°C). For every 5 mg of PAH add 2 mL of acetone and ensure that the PAH is completely dissolved, including any PAH that may be adhering to the walls of the container. For every 5 mg of PAH add 10 mL of a 0.3 M $KMnO_4$ solution in 3 M H_2SO_4 (freshly prepared) and swirl the mixture and allow to react for 1 h. The purple color should be maintained during this reaction time. If it is not, add more $KMnO_4$ solution until the reaction mixture remains purple for 1 h. At the end of the reaction decolorize the solution with ascorbic acid, neutralize, test for completeness of destruction, and discard it.
Note. Use *at least* 1 mL of acetone and *at least* 10 mL of $KMnO_4$.

Destruction of PAH in Agar

A. Cut the contents of the Petri dish into small pieces and homogenize with 30 mL of H_2O in a high speed blender. Extract the resulting solution twice with 30 mL portions of ethyl acetate (upper layer) and combine the extracts and dry them over anhydrous sodium sulfate. Remove the ethyl acetate by evaporation under reduced pressure using a rotary evaporator. For every 5 mg of PAH add 2 mL of acetone and ensure that the PAH is completely dissolved, including any PAH that may be adhering to the walls of the container. For every 5 mg of PAH add 10 mL of a 0.3 M KMnO$_4$ solution in 3 M H$_2$SO$_4$ (freshly prepared) and swirl the mixture and allow to react for 1 h. The purple color should be maintained during this reaction time. If it is not, add more KMnO$_4$ solution until the reaction mixture remains purple for 1 h. At the end of the reaction decolorize the solution with ascorbic acid, neutralize, test for completeness of destruction, and discard it.

Note. Use *at least* 1 mL of acetone and *at least* 10 mL of KMnO$_4$.

B. Cut the contents of the Petri dish into small pieces and homogenize with 30 mL of H$_2$O in a high speed blender. Extract the resulting solution twice with 30 mL portions of ethyl acetate (upper layer) and combine the extracts and dry them over anhydrous sodium sulfate. Remove the ethyl acetate by evaporation under reduced pressure using a rotary evaporator. For every 5 mg of PAH add 2 mL of DMSO and ensure that the PAH is completely dissolved, including any PAH that may be adhering to the walls of the container. For every 5 mg of PAH add 10 mL of concentrated H$_2$SO$_4$ (**Caution!** Exothermic process) and swirl the mixture and allow to react for at least 2 h. At the end of the reaction cautiously add the solution to at least three times the volume of cold H$_2$O (using an ice bath if desired as this is a very exothermic process), neutralize, test for completeness of destruction, and discard it.

Decontamination of Glassware Contaminated with PAH

A. Rinse the glassware with four portions of acetone which are large enough to wet the glassware thoroughly. Analyze the fourth rinse for the absence of PAH. Combine the rinses and remove the acetone by evaporation under reduced pressure using a rotary evaporator. For every 5 mg of PAH add 2 mL of acetone and ensure that the PAH is completely dissolved, including any PAH that may be adhering to the walls of the container. For every 5 mg of PAH add 10 mL of a 0.3 M KMnO$_4$ solution in 3 M H$_2$SO$_4$ (freshly prepared) and swirl the mixture and allow to react for 1 h. The purple color should be maintained during this reaction time. If it is not, add more KMnO$_4$ solution until the reaction mixture remains purple for 1 h. At the end of the reaction decolorize the solution with ascorbic acid, neutralize, test for completeness of destruction, and discard it.

Note. Use *at least* 1 mL of acetone and *at least* 10 mL of KMnO$_4$.

B. Add sufficient DMSO to wet the surface of the glass and then add five times this volume of concentrated H_2SO_4 (**Caution!** Exothermic process). Allow the mixture to react for 2 h with occasional swirling. At the end of the reaction cautiously add the solution to at least three times the volume of cold H_2O (using an ice bath if desired as this is a very exothermic process), neutralize, test for completeness of destruction, and discard it.

Treatment of Spills Involving PAH

Remove as much as possible of the spill by HEPA vacuuming or using absorbents. Treat the removed material using one of the methods described above. Wet the surface with DMF, then add a 0.3 M solution of $KMnO_4$ in 3 M H_2SO_4 and allow to react for 1 h. Take up the residual solution with an absorbent and test for completeness of destruction. Add more solvent to the spill area and absorb this with white paper. Examine the paper under long- and short-wavelength UV light for the presence of fluorescence attributable to PAH.

Analytical Procedures

High-pressure liquid chromatography methods of analysis for PAH have been reviewed.[22] Polycyclic aromatic hydrocarbons may also be analyzed by gas chromatography using a 1.8 m x 2-mm i.d. glass column packed with 3% OV-1 on 80/100 Supelcoport. The oven temperature is 260°C, the injection temperature is 300°C, and the temperature of the flame ionization detector is 300°C. Before analysis extract the neutralized reaction mixture three times with 10 mL portions of cyclohexane and dry the extracts over anhydrous sodium sulfate. Evaporate to dryness under reduced pressure using a rotary evaporator and take up the residue in 500 μL of acetone or other suitable solvent.

For 7-bromomethylbenz[a]anthracene, extract the reaction mixture three times with 10 mL portions of cyclohexane and dry the extracts over anhydrous sodium sulfate and evaporate. Take up the residue in 250 μL of toluene and spot 10 μL of this solution on a silica gel 60 thin-layer chromatography (TLC) plate. Develop the TLC plate with cyclohexane : ether (60:40) and then examine **immediately** under long-wavelength UV light. 7-Bromomethylbenz[a]anthracene has an R_f value of $\sim$ 0.28. Use of a standard solution of BrMBA is helpful for quantitation.

Mutagenicity Assays[1]

The reaction mixtures from these procedures were tested using *Salmonella typhimurium* strains TA98 and TA100, with and without activation. In general, they were not mutagenic.

Related Compounds

Although these procedures have only been validated for the compounds listed above, they shold be applicable to other PAH.

References

1. Castegnaro, M.; Grimmer,G.; Hutzinger,O.; Karcher,W.; Kunte, H.; Lafontaine,M.; Sansone, E. B.; Telling, G.; Tucker, S.P., Eds. *Laboratory Decontamination and Destruction of Carcinogens in Laboratory Wastes: Some Polycyclic Aromatic Hydrocarbons*; International Agency for Research on Cancer: Lyon, 1983 (IARC Scientific Publications No. 49).

2. Other names are 1,2-benzanthracene, 2,3-benzphenanthrene, benzanthrene, benzoanthracene, naphthanthracene, and tetraphene.

3. Other names are 3,4-benzpyrene, 1,2-benzpyrene, and benzo(*d,e,f*)chrysene.

4. Another name is 1,2:5,6-dibenzanthracene.

5. Other names are 9,10-dimethyl-1,2-benzanthracene and 1,4-dimethyl-2,3-benzphenanthrene.

6. Other names are 20-methylcholanthrene, 3-MECA, and 1,2-dihydro-3-methylbenz(*j*)aceanthrylene.

7. International Agency for Research on Cancer. *IARC Monographs on the Evaluation of the Carcinogenic Risk of Chemicals to Man*. Volume 3, *Certain Polycyclic Aromatic Hydrocarbons and Heterocyclic Compounds*; International Agency for Research on Cancer: Lyon, 1973; pp 45–68.

8. International Agency for Research on Cancer. *IARC Monographs on the Evaluation of the Carcinogenic Risk of Chemicals to Humans*. Volume 32, *Polynuclear Aromatic Compounds, Part 1, Chemical, Environmental and Experimental Data*; International Agency for Research on Cancer: Lyon, 1983; pp 135–145.

9. Reference 7, pp 91–136.

10. Reference 8, pp 211–224.

11. Dipple, A.; Levy, L.S.; Lawley, P.D. Comparative carcinogenicity of alkylating agents: comparisons of a series of alkyl and aralkyl bromides of differing chemical reactivities as inducers of sarcoma at the site of a single injection in the rat. *Carcinogenesis* **1981**, *2*, 103–107.

12. Reference 7, pp 178–196.

13. Reference 8, pp 299–308.

14. Griswold, D.P., Jr.; Casey, A.E.; Weisburger, E.K.; Weisburger, J.H.; Schabel, F.M. On the carcinogenicity of a single intragastric dose of hydrocarbons, nitrosamines, aromatic amines, dyes, coumarins, and miscellaneous chemicals in female Sprague-Dawley rats. *Cancer Res.* **1966**, *26*, 619–625.

15. Griswold, D.P., Jr.; Casey, A.E.; Weisburger, E.K.; Weisburger, J.H. The carcinogenicity of multiple intragastric doses of aromatic and heterocyclic nitro or amino derivatives in young female Sprague-Dawley rats. *Cancer Res.* **1968**, *28*, 924–933.

16. Rigdon, R.H. Pulmonary neoplasms produced by methylcholanthrene in the white Pekin duck. *Cancer Res.* **1961**, *21*, 571–574.

17. Blumenthal, H.T.; Rogers, J.B. Studies of guinea pig tumors. II. The induction of malignant tumors in guinea pigs by methylcholanthrene. *Cancer Res.* **1962**, *22*, 1155–1162.

18. Homburger, F.; Hsueh, S.-S.; Kerr, C.S.; Russfield, A.B. Inherited susceptibility of inbred strains of Syrian hamsters to induction of subcutaneous sarcomas and mammary and gastrointestinal carcinomas by subcutaneous and gastric administration of polynuclear hydrocarbons. *Cancer Res.* **1972**, *32*, 360–366.

19. Sax, N.I; Lewis, R.J., Sr. *Dangerous Properties of Industrial Materials*, 7th ed.; Van Nostrand-Reinhold: New York, 1989; pp 390–391.

20. Reference 19, pp 1362–1363.

21. Reference 19, pp 2286–2287.

22. Lee, M.L.; Novotny, M.V.; Bartle,K. *Analytical Chemistry of Polycyclic Aromatic Compounds*; Academic Press: New York, 1981.

β-PROPIOLACTONE

CAUTION! Refer to safety considerations section on page 6 before starting any of these procedures.

β-Propiolactone (BPL (**I**) bp 162°C) is a volatile liquid that is soluble to the extent of 37% in H_2O. It is used as a sterilant and industrially as an intermediate. It is an animal carcinogen.[1] Other names are 2-oxetanone, hydracrylic acid β-lactone, β-propionolactone, propanolide, Betaprone, and 3-hydroxypropionic acid lactone.

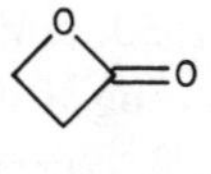

(I)

Principle of Destruction

β-Propiolactone is oxidized by potassium permanganate in the presence of 3 M sulfuric acid ($KMnO_4$ in H_2SO_4). The products have not been determined. Destruction was complete and <0.21% BPL remained.[2]

233

Destruction Procedure

Take up 0.5 mL (573 mg) of ß-propiolactone in 100 mL of 3 M H_2SO_4 and add 4.8 g of $KMnO_4$ in portions with stirring. Stir the reaction mixture at room temperature for 18 h, then decolorize it with ascorbic acid and neutralize with solid sodium bicarbonate. Test for completeness of destruction and discard it.

Analytical Procedures[2]

One hundred microliters of the solution to be analyzed was added to 1 mL of a solution of 2 mL of acetic acid in 98 mL of 2-methoxyethanol. This mixture was swirled and 1 mL of a solution of 5 g of 4-(4-nitrobenzyl)pyridine (4-NBP) in 100 mL of 2-methoxyethanol was added. The solution was heated at 100°C for 10 min, then cooled in ice for 5 min. Piperidine (0.5 mL) and 2-methoxyethanol (2 mL) were added and the violet color was determined at 560 nm using a UV/Vis spectrophotometer.

To check the efficacy of the analytical procedure, a small quantity of BPL can be added to the solution to be analyzed after the acetic acid-2-methoxyethanol has been added but before the 4-NBP is added. A positive response will indicate that the analytical technique is satisfactory. Using the analytical procedure described above with 10-mm disposable plastic cuvettes in a Gilford 240 UV/Vis spectrophotometer the limit of detection was 12 mg/L, but this can easily be reduced by using more than 100 μL of the reaction mixture.

Mutagenicity Assays

The mutagenicity assays were carried out as described on page 4 using tester strains TA98, TA100, TA1530, and TA1535. The final reaction mixture (tested at a level corresponding to 0.57 mg of undegraded BPL per plate) was not mutagenic. The pure compound was highly mutagenic.[2]

Related Compounds

This is a general oxidative procedure and should work for other lactones although it would have to be validated for each compound.

References

1. International Agency for Research on Cancer. *IARC Monographs on the Evaluation of Carcinogenic Risk of Chemicals to Man.* Volume 4, *Some Aromatic Amines, Hydrazine and Related Substances,* N-*Nitroso Compounds and Miscellaneous Alkylating Agents*; International Agency for Research on Cancer: Lyon, 1974, pp 259–269.

2. Lunn, G.; Sansone, E.B. Validated methods for degrading hazardous chemicals: Some alkylating agents and other compounds. *J. Chem. Educ.* (in press).

SODIUM AMIDE

CAUTION! Refer to safety considerations section on page 6 before starting any of these procedures.

Sodium amide ($NaNH_2$, sodamide) is a crystalline solid, which can react explosively with H_2O, with heat, or on grinding. It may become explosive on storage.[1] Sodium amide is used in organic synthesis.

Destruction Procedure[2]

Immerse the compound in toluene and slowly add 95% ethanol with stirring. The $NaNH_2$ is converted to sodium ethoxide and ammonia. When the reaction is complete dilute the reaction mixture with H_2O and discard it. Wash out contaminated apparatus with ethanol before cleaning.

Related Compounds

This procedure is also applicable to potassium amide.

237

References

1. Sax, N.I; Lewis, R.J., Sr. *Dangerous Properties of Industrial Materials*, 7th ed.; Van Nostrand-Reinhold: New York, 1989; p. 3041.

2. Bergstrom, F.W. Sodium amide. In *Organic Syntheses*; Horning, E.C., Ed.; Wiley: New York, 1955; Coll. Vol. 3, pp 778–783.

6-THIOGUANINE AND 6-MERCAPTOPURINE

> **CAUTION!** Refer to safety considerations section on page 6 before starting any of these procedures.

The degradation of a number of antineoplastic drugs, including 6-thioguanine and 6-mercaptopurine, was investigated by the International Agency for Research on Cancer (IARC).[1] 6-Thioguanine (mp >360°C) (**I**)[2] and 6-mercaptopurine (mp 313–314°C) (**II**)[3] are solids; they are insoluble in H_2O and organic solvents but soluble in dilute acid or base. 6-Mercaptopurine is mutagenic.[4] 6-Thioguanine[5] and 6-mercaptopurine[4] are teratogens. These compounds are employed as antineoplastic drugs.

(I)   **(II)**

Principle of Destruction

6-Thioguanine and 6-mercaptopurine are destroyed by oxidation with potassium permanganate in sulfuric acid ($KMnO_4$ in H_2SO_4).[1] Destruction is >99.5%.

Destruction Procedures

Destruction of Bulk Quantities

Dissolve in 3 M H_2SO_4 so that the concentration does not exceed 0.9 mg/mL, then add 0.5 g of $KMnO_4$ for each 80 mL of solution and stir overnight. Decolorize with ascorbic acid, neutralize, check for completeness of destruction, and discard it.

Destruction of Aqueous Solutions

Dilute with H_2O, if necessary, so that the concentration does not exceed 0.9 mg/mL, then add enough concentrated H_2SO_4 to obtain a 3 M solution and allow it to cool to room temperature. For each 80 mL of solution add 0.5 g of $KMnO_4$ and stir overnight. Decolorize with ascorbic acid, neutralize, check for completeness of destruction, and discard it.

Destruction of Pharmaceutical Preparations

To solutions of 7.5 mg of 6-thioguanine in 50 mL of 5% dextrose solution or 10 mg of 6-mercaptopurine in 10 mL of 5% dextrose solution add enough concentrated H_2SO_4 to obtain a 3 M solution. Allow the solution to cool to room temperature. Dissolve oral preparations of these drugs in 3 M H_2SO_4. For each 80 mL of any of these solutions add 4 g of $KMnO_4$, in small portions to avoid frothing, and stir the mixture overnight. Decolorize with ascorbic acid, neutralize, check for completeness of destruction, and discard it.

Destruction of Solutions in Volatile Organic Solvents

Remove the solvent under reduced pressure using a rotary evaporator and take up the residue in 3 M H_2SO_4, so that the concentration of the drug does not exceed 0.9 mg/mL. For each 80 mL of solution add 0.5 g of

$KMnO_4$ and stir overnight. Decolorize with ascorbic acid, neutralize, check for completeness of destruction, and discard it.

Destruction of Dimethyl Sulfoxide (DMSO) or Dimethylformamide (DMF) Solutions

Dilute with H_2O so that the concentration of DMSO or DMF does not exceed 20% and the concentration of the drug does not exceed 0.9 mg/mL, then add enough concentrated H_2SO_4 to obtain a 3 M solution and allow it to cool to room temperature. For each 80 mL of solution add 4 g of $KMnO_4$, in small portions to avoid frothing, and stir overnight. Decolorize with ascorbic acid, neutralize, check for completeness of destruction, and discard it.

Decontamination of Glassware

Immerse the glassware in a 0.3 M solution of $KMnO_4$ in 3 M H_2SO_4 for 10–12 h, then decolorize it by immersion in ascorbic acid solution.

Decontamination of Spills

Allow any organic solvent to evaporate and remove as much of the spill as possible by HEPA vacuuming (not sweeping), then rinse the area with 0.1 M H_2SO_4. Take up the rinse with absorbents and allow the rinse and absorbents to react with 0.3 M $KMnO_4$ solution in 3 M H_2SO_4 overnight. If the color fades, add more solution. Check for completeness of decontamination by using a wipe moistened with 0.1 M sodium hydroxide solution. Analyze the wipe for the presence of the drug.

Analytical Procedures

These drugs can be analyzed by HPLC using a 25-cm reverse phase column and UV detection at 340 nm. The mobile phases which have been recommended are as follows:

6-Thioguanine	0.02 M KH_2PO_4 : acetonitrile (98:2) flowing at 1.5 mL/min
	or 0.1 mM KH_2PO_4 : methanol (92.5:7.5) flowing at 1 mL/min

6-Mercaptopurine   0.02 *M* KH$_2$PO$_4$: acetonitrile (98:2) flowing at 1.5
                    mL/min
                    or 0.1 m*M* KH$_2$PO$_4$ flowing at 1 mL/min

Mutagenicity Assays

In the IARC study[1] tester strains TA98, TA100, and TA1535 of *Salmonella typhimurium* were used with and without mutagenic activation. The reaction mixtures were not mutagenic.

Related Compounds

Potassium permanganate in H$_2$SO$_4$ is a general oxidative method and should, in principle, be applicable to many drugs. However, any new application should be thoroughly validated both for complete destruction of the compound and for the production of nonmutagenic reaction mixtures.

References

1. Castegnaro, M.; Adams, J.; Armour, M-.A.; Barek, J.; Benvenuto, J.; Confalonieri, C.; Goff, U.; Ludeman, S.; Reed, D.; Sansone, E. B.; Telling, G., Eds. *Laboratory Decontamination and Destruction of Carcinogens in Laboratory Wastes: Some Antineoplastic Agents*; International Agency for Research on Cancer: Lyon, 1985 (IARC Scientific Publications No. 73).

2. Other names are 2-amino-1,7-dihydro-6*H*-purine-6-thione, Lanvis, 2-aminopurine-6-thiol, 2-aminopurine-6(1*H*)-thione, 2-amino-6-mercaptopurine, and Tabloid.

3. Other names are purine-6-thiol, 7-mercapto-1,3,4,6-tetrazaindene, 6MP, Leukerin, Mercaleukin, Purinethol, and 6-purinethiol.

4. Sax, N.I; Lewis, R.J., Sr. *Dangerous Properties of Industrial Materials*, 7th ed.; Van Nostrand-Reinhold: New York, 1989; pp 2921–2922.

5. Reference 4, pp 218–219.

APPENDIX

Recommendations for Wipe Solvents for Use After Spill Cleanup

After a chemical spill has been cleaned up and the area decontaminated it is frequently helpful to use a moistened wipe to take a sample of the decontaminated surface to ensure that decontamination is complete. The wipe should be moistened with a reagent that will dissolve the spilled compound and will not interfere with the subsequent analysis. This list has been compiled on the basis of information in the published literature and a few tests. The use of a wipe sample is generally not appropriate with highly reactive compounds and these cases are indicated by NA. In some cases a different reagent may be required depending on the circumstances of the spill and the cleanup. These instances are detailed in the footnotes.

We list here the names of compounds as used in this book. Please consult the Molecular Formula Index, the CAS Registry Number Index, or the Name Index for compounds that may be known by alternate names.

Acetic anhydride	NA
Acetonitrile	NA
Acetyl chloride	NA
Acid anhydrides	NA
Acid chlorides	NA
Acid halides	NA
Acyl chlorides	NA
Acyl halides	NA

Adriamycin	Water
Aflatoxin B_1	Methanol
Aflatoxin B_2	Methanol
Aflatoxin G_1	Methanol
Aflatoxin G_2	Methanol
Aflatoxin M_1	Methanol
Aflatoxins	Methanol
Alkali metals	NA
Alkylating agents	See individual compounds
4-Aminobiphenyl	Methanol[1]
N-Aminomorpholine	Methanol[1]
N-Aminopiperidine	Methanol[1]
N-Aminopyrrolidine	Methanol[1]
Ammonium dichromate	Water
Anhydrides	NA
Antineoplastic drugs	See individual compounds
Aromatic amines	See individual compounds
Azides	See individual compounds
Azobenzene	Methanol[1]
Azocompounds	See individual compounds
4,4'-Azoxyanisole	Methanol
Azoxybenzene	Methanol
Azoxycompounds	See individual compounds
Azoxymethane	Methanol
BCME	Methanol
BCNU	Methanol
Benz[a]anthracene	Acetone
Benzal chloride	Methanol
Benzenesulfonyl chloride	Methanol
Benzidine	Methanol[1]
Benzo[a]pyrene	Acetone
Benzonitrile	Methanol
Benzoyl chloride	NA
Benzoyl peroxide	Acetone
Benzyl azide	Methanol
Benzyl bromide	Methanol
Benzyl chloride	Methanol
Benzyl cyanide	Methanol
Bis(chloromethyl)ether	Methanol

Boron trifluoride	NA
Bromobenzene	Methanol
4-Bromobenzoic acid	Methanol[2]
1-Bromobutane	Methanol
2-Bromobutane	Methanol
1-Bromodecane	Methanol
2-Bromoethanol	Water
2-Bromoethylamine	Water
7-Bromomethylbenz[a]anthracene	Acetone
2-Bromo-2-methylpropane	Methanol
4-Bromophenylacetic acid	Methanol[2]
Butadiene diepoxide	Water
n-Butyl bromide	Methanol
s-Butyl bromide	Methanol
t-Butyl bromide	Methanol
n-Butyl chloride	Methanol
n-Butyl iodide	Methanol
s-Butyl iodide	Methanol
t-Butyl iodide	Methanol
t-Butyl peroxide	Methanol
Calcium carbide	NA
Calcium hydride	NA
Carbamic acid esters	Water
Carbon disulfide	NA
Carmustine	Methanol
CCNU	Methanol
Chlorambucil	Methanol
Chloroacetic acid	Water
2- and 3-Chloroaniline	Methanol[1]
Chlorobenzene	Methanol
2-Chlorobenzoic acid	Methanol[2]
1-Chlorobutane	Methanol
1-Chlorodecane	Methanol
2-Chloroethanol	Water
2-Chloroethylamine	Water
2-Chloro-5-fluorobenzoic acid	Methanol[2]
2-Chlorohydroquinone	Methanol
2-Chloroisophthalic acid	Methanol[2]
Chloromethylmethylether	Methanol

4-(4-Chloro-2-methylphenyl)butyric acid	Methanol[2]
4-(4-Chloro-3-methylphenyl)butyric acid	Methanol[2]
Chloromethylsilanes	NA
2-, 3-, and 4-Chloronitrobenzene	Methanol
4-Chlorophenol	Methanol
4-Chlorophenoxyacetic acid	Methanol[2]
2-Chlorophenylacetic acid	Methanol[2]
2-Chlorophenyl mercaptan	Methanol
3-Chloropyridine	Methanol
2-Chlorothiophenol	Methanol
Chlorotrimethylsilane	NA
Chlorozotocin	Methanol
Chromerge	Water
Chromic acid	Water
Chromium trioxide	Water
Chromium(VI)	Water
Cisplatin	Water
CMME	Methanol
Complex metal hydrides	NA
Cyanogen bromide	Water
Cyanogen chloride	Water
Cyanogen iodide	Water
Cyclophosphamide	Water
Cycloserine	Water
Dacarbazine	0.1 M HCl[3]
Daunomycin	Water
Daunorubicin	Water
Dialkyl sulfates	See individual compounds
Diaminobenzidine	Methanol[1]
Di(4-amino-3-chlorophenyl)methane	Methanol[1]
N,N'-Diaminopiperazine	Methanol[1]
2,4-Diaminotoluene	Methanol[1]
Dibenz[a,h]anthracene	Acetone
1,1-Dibutylhydrazine	Methanol[1]
3,3'-Dichlorobenzidine	Methanol[1]
2,4-Dichlorobenzoic acid	Methanol[2]
3,4-Dichlorobenzoic acid	Methanol[2]
Dichlorodimethylsilane	NA
Dichloromethotrexate	0.1 M HCl[3]

2,4-Dichlorophenylacetic acid	Methanol[2]
3,4-Dichlorophenylacetic acid	Methanol[2]
α,α-Dichlorotoluene	Methanol
1,1-Diethylhydrazine	Methanol[1]
Diethyl sulfate	Methanol
1,1-Diisopropylhydrazine	Methanol[1]
Diisopropylnitramine	Methanol
3,3'-Dimethoxybenzidine	Methanol[1]
N,N-Dimethyl-4-amino-4'-hydroxyazobenzene	Methanol[1]
7,12-Dimethylbenz[a]anthracene	Acetone
3,3'-Dimethylbenzidine	Methanol[1]
Dimethyl disulfide	Methanol
1,1-Dimethylhydrazine	Water
1,2-Dimethylhydrazine	Water
Dimethyl sulfate	Methanol
N,N-Dinitrosopiperazine	Methanol
1,5-Diphenylcarbazide	Methanol
Diphenyl disulfide	Acetone
1,2-Diphenylhydrazine	Methanol[1]
Disulfides	See individual compounds
Doxorubicin	Water
Drugs	See individual compounds
ENNG	Methanol
ENU	Methanol
ENUT	Methanol
Erythritol anhydride	Water
Ethanethiol	Water
Ethidium bromide	Water
3-Ethoxypropionitrile	Water
Ethyl carbamate	Water
Ethyl mercaptan	Water
Ethyl methanesulfonate	Water
N-Ethyl-N'-nitro-N-nitrosoguanidine	Methanol
N-Ethyl-N-nitrosourea	Methanol
N-Ethyl-N-nitrosourethane	Methanol
N-Ethylurethane	Water
4-Fluorobenzoic acid	Methanol[2]
Haloethers	See individual compounds
Homidium bromide	Water

Hydrazine	Water
Hydrazobenzene	Methanol[1]
Hydrides, complex metal	NA
Hydrogen cyanide	NA
Ifosfamide	Water
Inorganic cyanides	Water
Inorganic fluorides	Water
Iodobenzene	Methanol
1-Iodobutane	Methanol
2-Iodobutane	Methanol
Iodomethane	Methanol
2-Iodo-2-methylpropane	Methanol
Iproniazid phosphate	Water
Isoniazid	Water
Isophosphamide	Water
LAH	NA
Lithium	NA
Lithium aluminum hydride	NA
Lomustine	Methanol
Mechlorethamine	Water
Melphalan	Methanol
Mercaptans	See individual compounds
6-Mercaptopurine	0.1 M KOH[3]
Metal hydrides	NA
Methanesulfonates	See individual compounds
Methanethiol	NA
Methotrexate	0.1 M HCl[3]
2-Methylaziridine	Water
Methyl carbamate	Water
Methyl CCNU	Methanol
3-Methylcholanthrene	Acetone
Methyl disulfide	Methanol
Methylhydrazine	Methanol[1]
Methyl iodide	Methanol
Methyl mercaptan	NA
Methyl methanesulfonate	Methanol
N-Methyl-N'-nitro-N-nitrosoguanidine	Methanol
N-Methyl-N-nitrosoacetamide	Methanol
N-Methyl-N-nitroso-p-toluenesulfonamide	Methanol

N-Methyl-N-nitrosourea	Methanol
N-Methyl-N-nitrosourethane	Methanol
1-Methyl-1-phenylhydrazine	Methanol[1]
1-Methyl-4-phenyl-1,2,3,6-tetrahydropyridine	Methanol[1]
3-Methyl-1-p-tolyltriazene	Methanol[1]
Methyltrichlorosilane	NA
N-Methylurethane	Water
Mitomycin C	Methanol
MNNG	Methanol
MNTS	Methanol
MNU	Methanol
MNUT	Methanol
MOCA	Methanol[1]
MPTP	Methanol[1]
1-Naphthylamine	Methanol[1]
2-Naphthylamine	Methanol[1]
4-Nitrobiphenyl	Methanol
Nitrogen mustards	See individual compounds
N-Nitromorpholine	Methanol
Nitrosamides	See individual compounds
Nitrosamines	See individual compounds
N-Nitrosodibutylamine	Methanol
N-Nitrosodiethylamine	Water
N-Nitrosodiisopropylamine	Methanol
N-Nitrosodimethylamine	Water
N-Nitrosodiphenylamine	Methanol
N-Nitrosodipropylamine	Methanol
N-Nitroso-N-methylaniline	Methanol
N-Nitrosomorpholine	Water
N-Nitrosopiperidine	Water
N-Nitrosopyrrolidine	Water
Nitrosoureas	See individual compounds
Organic azides	See individual compounds
Organic bromides	See individual compounds
Organic chlorides	See individual compounds
Organic cyanides	See individual compounds
Organic fluorides	See individual compounds
Organic halides	See individual compounds
Organic nitriles	See individual compounds

Osmium tetroxide	Footnote[4]
PCNU	Methanol
Peroxides	See individual compounds
Phenyl azide	Methanol
4-Phenylazoaniline	Methanol[1]
4-Phenylazophenol	Methanol[1]
Phenyl cyanide	Methanol
Phenyl disulfide	Acetone
Phenylhydrazine	Methanol
Phenyl mercaptan	Methanol
Phosphorus	NA
Phosphorus pentoxide	NA
Picric acid	Water
Polycyclic aromatic hydrocarbons	See individual compounds
Potassium	NA
Potassium amide	NA
Potassium cyanide	Water
Potassium dichromate	Water
Potassium fluoride	Water
Procarbazine hydrochloride	Water
1,3-Propane sultone	Water
Propidium iodide	Water
β-Propiolactone	Water
Propyleneimine	Water
Sodium	NA
Sodium amide	NA
Sodium azide	Water
Sodium borohydride	NA
Sodium cyanide	Water
Sodium dichromate	Water
Sodium fluoride	Water
Sodium hydride	NA
Spirohydantoin mustard	Methanol
Streptozotocin	Water
Sulfides, inorganic	Water
Sulfonyl chlorides	NA
Sulfonyl halides	NA
Tetramethyltetrazene	Methanol[1]
Tetrazenes	See individual compounds

6-Thioguanine	0.1 M KOH[3]
Thiols	See individual compounds
Thiophenol	Methanol
p-Toluenesulfonyl chloride	Methanol
p-Tolylhydrazine hydrochloride	Water
Triazenes	See individual compounds
2,4,6-Tribromophenol	Methanol
2,2,2-Trichloroacetic acid	Water[2]
2,4,6-Trichlorophenol	Methanol
Trimethylsilyl chloride	NA
2,4,6-Trinitrophenol	Water
Uracil mustard	Methanol
Urethane	Water
Vinblastine sulfate	Water
Vincristine sulfate	Water

Footnotes

1. If the compound is present as the hydrochloride or other salt or if the decontamination procedure produces the hydrochloride or other salt, water should be used instead of methanol. Neutralization of this solution may be necessary before analysis.

2. If the compound is present as the sodium or other salt or if the decontamination procedure produces the sodium or other salt, water should be used instead of methanol. Neutralization of this solution may be necessary before analysis.

3. Neutralization of this solution may be necessary before analysis.

4. Osmium tetroxide is soluble in water but a better method for determining completeness of decontamination may be to suspend either a glass cover slip coated in corn oil (Mazola) or a piece of filter paper soaked in corn oil over the area. Blackening indicates that osmium tetroxide is still present.

MOLECULAR FORMULA INDEX

Compounds which generally come as the hydrochloride or other salt forms are listed only as the free bases. Water of hydration is ommitted. Inorganics are listed first.

253

CAS REGISTRY NUMBER INDEX

Some compounds may have more than one Registry Number depending on whether the free base, salt, D form, L form, etc. is considered. We have tried to include as many of these Registry Numbers as possible but in each case we give only the name used in the monographs.

NAME INDEX

Although we have tried to include as many synonyms as possible, this list is not exhaustive. In particular we have not incorporated all possible minor variations, e.g., *p*-methyl as well as 4-methyl, and so the user should check all common variants. Note that we have used 2-chloroethyl instead of β-chloroethyl throughout and we have not included D, L, DL, or *meso* forms. Compounds which generally come as the hydrochloride or other salt forms are listed only as the free bases. The page numbers given are the first page of the monograph to emphasize the point that the entire chapter should be read before commencing any work. The compound in question, however, may be listed further down in the Related Compounds section or in the footnotes where synonyms are sometimes given.